Ecotones

CONTRIBUTORS

David L. Correll
Smithsonian Environmental
Research Center
647 Contees Wharf Rd.
Box 28
Edgewater, MD 21037

Henri Décamps
Centre National de la Récherche
Scientifique
29, rue J. Marvig
31055 Toulouse
France

Robert H. Gardner
Environmental Sciences Division
P.O. Box 2008
Oak Ridge National Laboratory
Oak Ridge, TN 37831

James R. Gosz
Department of Biology
University of New Mexico
Albuquerque, NM 87131

Stanley V. Gregory
Department of Fisheries and Wildlife
Oregon State University
Corvallis, OR 97331

Marjorie M. Holland
Public Affairs Office
Ecological Society of America
9650 Rockville Pike, Suite 2503
Bethesda, MD 20814

Sandra Lavorel
Centre d'Ecologie Fonctionelle
et Evolutive
Centre National de la Récherche
Scientifique
Route de Mende
34033 Montpellier
France

Robert Naiman
Center for Streamside Studies, AR-10
University of Washington
Seattle, WA 98195

Ronald P. Neilson
Environmental Protection Agency
Oregon State University
200 S.W. 35th St.
Corvallis, OR 97333

Robert V. O'Neill
Environmental Sciences Division
P.O. Box 2008
Oak Ridge National Laboratory
Oak Ridge, TN 37831

Henry Regier
Institute for Environmental Studies
University of Toronto
Toronto, Ontario M5S 1A1
Canada

Paul G. Risser
Scholes Hall Rm. 108
Univeristy of New Mexico
Albuquerque, NM 87131

James Sedell
Forestry Sciences Laboratory
3200 SW Jefferson Way
Corvallis, OR 97331

Robert J. Steedman
Ontario Ministry of Natural Resources
955 Oliver Rd.
Thunder Bay, Ontario P7B 5E1
Canada

Monica G. Turner
Environmental Sciences Division
P.O. Box 2008
Oak Ridge National Laboratory
Oak Ridge, TN 37831

Ecotones

The Role of Landscape Boundaries in the Management and Restoration of Changing Environments

Edited by

Marjorie M. Holland
Paul G. Risser
Robert J. Naiman

Chapman
and Hall

New York

London

First published in 1991 by
Chapman & Hall
an imprint of
Routledge, Chapman & Hall, Inc.
29 West 35th Street
New York, NY 10001

Published in Great Britain by
Chapman & Hall
2-6 Boundary Row
London SE1 8HN

Printed in the United States of America

COVER PHOTOGRAPH: SPOT satellite photograph of the Ganges River delta, India

Published with the support of the Ecological Society of America and the U.S. Man in Biosphere Program.

Library of Congress Cataloguing in Publication Data

Ecotones: the role of landscape boundaries in the management and restoration of changing environments/edited by Marjorie M. Holland and Paul G. Risser and Robert J. Naiman.
 p. cm.
 Papers presented at a symposium held at the annual meeting of the Ecological Society of America in Toronto, August 8, 1989.
 Includes bibliographical references and index.
 ISBN 0-412-03091-8
 1. Restoration ecology—Congresses. 2. Ecotones—Congresses. I. Holland, Marjorie.
II. Risser, Paul G. III. Naiman, Robert J. IV. Ecological Society of America. Meeting (1989: Toronto, Ont.)
QH541.15.R45E27 1991
574.5'26—dc20   91-19208
                                                                                                          CIP

British Library Cataloguing in Publication Data

Holland, Marjorie
 Ecotones: the role of landscape boundaries in the management and restoration of
 changing environments. I. Title II. Risser, Paul G. (Paul Gillan), 1939-III. Naiman, Robert J.
 574.524

 ISBN 0412030918

TABLE OF CONTENTS

EDITORS' FOREWORD

We live in a changing world; one in which there is much concern and discussion about the topics of global change, loss of biodiversity, and increasing threats to the sustainability of ecosystems. The effects these changes may have on the environment have lead governments and scientists to make predictions as to how soon changes might occur, where, and with what impact for large and small regions of the Earth. Along with this concern for change in various regions has come the need to understand the role of boundaries between these regions and between landscape elements.

Much previous ecological research has dealt with processes within relatively homogeneous landscape units or even the collective characteristics of a composite landscape. Now, however, there is an appreciation that abiotic and biotic components move across heterogeneous landscapes and that the boundaries between these units take on important control functions in this dynamic spatial system. Furthermore, landscape boundaries (or ecotones) are important not only in satisfying life-cycle needs of many organisms, but generally are characterized by high biological diversity.

Three years ago the editors recognized the need for greater understanding of ecotones, and agreed to convene a meeting of recognized experts to discuss the role of landscape boundaries in the management and restoration of changing environments. Funding was secured in February 1988 by members of the Ecological Society of America (ESA)'s Public Affairs Committee and Public Affairs Office to develop an International Symposium in cooperation with the United States Man and the Biosphere (MAB) Program. It was with the support and encouragement of ESA's Public Affairs and Executive Committee members that this Symposium was held at the University of Toronto in August of 1989.

There were three major scientific events occurring in 1988 that made plans for our symposium particularly compelling. First, U.S. MAB had recently reorganized around five themes articulating human-natural system interactions, including the unique opportunities in the international network of Biosphere Reserves. Second, the International Geosphere-Biosphere Program (IGBP) was just beginning the organizational phases of a ten year, major research program on global processes. This large program is international in scope and focuses on geosphere-biosphere phenomena of time scales from decades to centuries. Finally, and most importantly, there was a widespread renewed attention toward the need to understand the environmental processes occurring in the boundaries between landscape units.

The 1989 symposium was the first opportunity for a large segment of the North American scientific community to discuss a developing, international, scientific interest in these boundaries. The symposium itself, building on two small international conferences in 1988, included presentations on the characteristics of ecotones, how ecotones respond to environmental change, and the ways in which ecotones can be most appropriately managed. This symposium served as a comprehensive planning tool for the North American research community, and encouraged the ecological community to play a catalytic role in shaping research efforts on landscape boundaries.

Many individuals deserve recognition for their efforts in making this volume a success. The dedication and persistence of each of the contributing authors is gratefully acknowledged. We thank Cecile Ledsky and Roger Soles of the United States Man and the Biosphere Program for their support and enthusiasm for this project. Financial assistance was provided by the United States Man and the Biosphere Program (U.S. State Department) and by the Ecological Society of America. All of the articles in this volume were evaluated by anonymous reviewers. Production of the volume would not have been accomplished without the advice and expertise of Raymond Prach. In addition, we appreciate the interest and cooperation we have received from Gregory Payne, Editor at Chapman and Hall.

Finally, we are especially grateful to staff of the Ecological Society of America's Public Affairs Office who devoted long hours and considerable talent to editing, proofreading, and formatting the papers which follow. We thank Yaffa Grossman, Mari Jensen, Tara Fuad, and Alfreda Thomas: their assistance has been invaluable and greatly appreciated.

Marjorie M. Holland
Bethesda, Maryland

Paul G. Risser
Albuquerque, New Mexico

Robert J. Naiman
Seattle, Washington

22 April 1991

THE ROLE OF LANDSCAPE BOUNDARIES IN THE MANAGEMENT AND RESTORATION OF CHANGING ENVIRONMENTS: INTRODUCTION

MARJORIE M. HOLLAND AND PAUL G. RISSER. The Ecological Society of America, Public Affairs Office, 9650 Rockville Pike, Bethesda, Maryland 20814, and University of New Mexico, Scholes Hall 108, Albuquerque, New Mexico 87131, USA.

Abstract.   Much previous ecological research has dealt with processes within relatively homogeneous landscape units or even the collective characteristics of a composite landscape. Now, however, there is an appreciation that abiotic and biotic components move through heterogeneous landscapes and that the boundaries between these landscape units take on important control functions in such dynamic spatial systems.   Also, landscape boundaries (ecotones) are important for satisfying life-cycle needs of many organisms, and generally are characterized by high biological diversity. This symposium, building on two small conferences in 1988, includes presentations on the characteristics of ecotones, how ecotones respond to environmental change, and the ways in which ecotones can be most appropriately managed.

Key words: ecotone, landscape boundary, global change, MAB, SCOPE, IGBP.

BACKGROUND

The ecotone concept was used in 1905 by Clements to denote the junction zone between two communities, where the processes of exchange or competition between neighboring formations might be readily observed. More recently, E.P. Odum (1971) defined an ecotone as: "a transition between two or more diverse communities as, for example, between forest and grassland or between a soft bottom and hard bottom marine community. . . . The ecotonal community commonly contains many of the organisms of each of the overlapping communities and, in addition, organisms which are characteristic of and often restricted to the ecotone. . . . This tendency for increased variety and density at community junctions is known as the edge effect."

Renewed European interest in the ecotone concept prompted us to convene a North American forum within which we could illustrate the relevance of ecotones to current ecological research. Much previous ecological research has dealt with processes within relatively homogeneous landscape units or the collective characteristics of a composite landscape. Now, however, there is an appreciation that abiotic and biotic components must move across heterogeneous landscapes and that, therefore, boundaries between units take on important control functions in such a dynamic spatial system.   For example, boundaries operate as controls or filters to nutrients, sediment, and water. Ecotones are important for satisfying life-cycle needs of many organisms, and are generally characterized by high biological diversity. Because landscape boundaries have such intense biotic-abiotic interactions, they may be particularly sensitive indicators of change in the global environment.

PURPOSE OF PAPER

An increasing number of scientists now believe that a study of ecotones will be of both theoretical and practical value (Risser 1985, Holland 1988, Naiman et al. 1988a). Ecotones harbor rich assemblages of flora and fauna, and serve as controls for the movement of water and materials throughout the landscape (Holland 1988). Because of these characteristics, using ecotones to prudently manage the biosphere is receiving renewed interest and support.

This paper describes the intellectual and logistical development of a scientific program on research and management of landscape boundaries. We present findings from two small conferences on this topic held in 1988, one sponsored by Unesco's Man and the Biosphere (MAB) Program and one by the International Council of Scientific Unions (ICSU)'s Scientific Committee on Problems of the Environment (SCOPE). MAB's Project Area 5, which addresses ecological processes in various freshwater and coastal aquatic ecosystems, is developing a research program on ecotones to be undertaken from now until 1996 (Naiman et al. 1988*b*, Naiman et al. 1989). The SCOPE project on ecotones is examining patterns and processes of biodiversity and ecological flows associated with the management of ecotones and their response to climate change (Hansen et al. 1988, di Castri and Hansen *in press*).

All papers in this volume were presented at a Symposium on "The role of landscape boundaries in the management and restoration of changing environments" held at the annual meeting of The Ecological Society of America (ESA) in Toronto, Canada, on 8 August 1989.

ECOTONES IN RELATION TO GLOBAL CHANGE

The international scientific community, in response to the overwhelming evidence that the earth system is changing in ways that are not fully understood, is embarking on a long-term research program. The International Geosphere-Biosphere Program (IGBP), launched by ICSU in 1986, aims to describe and understand the interactive physical, chemical, and biological processes that regulate the total earth system, the unique environment it provides for life, the changes that are occurring in that system, and the manner by which these changes are influenced by human activities (DeFries and Malone 1989).

The United Nation's World Commission on Environment and Development, under the leadership of Dr. Gro Harlem Brundtland, former Prime Minister of Norway, in 1987 addressed the broad spectrum of environmental, social, economic, and political issues associated with "sustainable development" (Brundtland 1987). The release of Our Common Future, the report produced by the World Commission in 1987, increased

political interest in issues of global environment and sustainable development all over the world.

In response to the growing international awareness of the potentially serious consequences of global climate change, the governing bodies of the United Nations Environment Program (UNEP) and the World Meteorological Organization (WMO) established the Intergovernmental Panel on Climate Change (IPCC). The IPCC is designed to serve as the primary international forum for assessing the state of knowledge about global climate change and its impacts, and considering possible response strategies.

Within the United States, a federal interagency committee and a National Academy of Science Committee have together successfully launched an interdisciplinary program to establish a sound scientific basis for developing national and international policy on global change issues. The overall U.S. strategy to address global change requires efforts in three areas: research to understand the Earth's environment; research and development of new technologies to adapt to, or mitigate, environmental changes; and formulation of the national and international policy response options needed to cope with a changing environment (Committee on Earth Sciences 1989).

Early in 1989, the Office of Science and Technology Policy (OSTP) transmitted to the U.S. Congress a report which accompanied the President's Fiscal Year 1990 Budget outlining the goals, implementation strategy and research budget of the U.S. Global Change Research Program. This strategy document, entitled "Our Changing Planet: A U.S. Strategy for Global Change Research," was the product of an intense interagency effort by experts in various earth sciences and other disciplines. This interagency effort was coordinated by the Committee on Earth Sciences (CES) of the Federal Coordinating Council for Science, Engineering, and Technology.

The scientific objectives of the U.S. research plan are to monitor, understand, and ultimately predict the causes and consequences of global change. It outlines a priority framework for focusing and integrating the interagency research efforts to ensure that they meet these objectives. This priority framework was derived from numerous research priorities outlined by both the U.S. and international communities.

The IGBP is now beginning the organizational phases of a ten year, major research program on global processes. This large international program will focus on geosphere-biosphere phenomena of time scales from decades to centuries. Measurement of landscape boundaries and boundary changes during alteration in the global environment have been identified as important subjects of research throughout the IGBP planning.

INTERNATIONAL MEETINGS ON ECOTONES

During the last four years, five international meetings have articulated current and future needs for understanding the dynamic processes occurring at landscape boundaries (di Castri et al. 1988).

At the first of these meetings, in January 1987, a new definition of an ecotone was developed, and seven critical questions were identified. Jointly sponsored by SCOPE and MAB, a working group of eleven scientists defined an ecotone as a "zone of transition between adjacent ecological systems, having a set of characteristics uniquely defined by space and time scales and by the strength of the interactions between adjacent ecological systems" (Holland 1988). This working definition is different from that of Odum (1971) in its consideration of space and time scales as well as in its attention to "the strength of the interactions between adjacent ecological systems." This relatively general definition has served as a launching point to develop the theoretical base necessary for future discussions (by ESA, SCOPE, MAB, and IGBP)

of the ecotone concept (Holland 1988).

In addition, that meeting identified seven preliminary research questions (Holland and Hansen 1988):

1. To what extent can ecotones be functionally classified so as to facilitate comparisons among different ecotones with respect to origin, structure and ecological processes?

2. Do ecotones provide stability for the resource patches they separate and, if so, at what spatial and temporal hierarchical scales do they operate?

3. What are the key attributes (processes and components) of ecotones that impart resistance and resilience to the resource patches adjacent to disturbance?

4. Is there a predictable pattern to dynamic change in ecotones under natural conditions?

5. What are the characteristics and processes of ecotones that are sensitive to changes in the global environment?

6. What is the importance of ecotones in maintaining local, regional and global biodiversity?

7. How have humans maintained and restored ecotones in the past? Might that level be expected to continue, diminish or intensify in the future? At what spatial and temporal scales are research results most useful for decision-making and management?

A second planning meeting in mid-1987 included scientists affiliated with MAB or with Unesco's International Hydrological Program (IHP). This meeting identified elements for a program on management options for the conservation and restoration of land/inland water ecotones through increased understanding of ecological processes (Naiman et al. 1988b). In developing a new generation of MAB field activities for investigating the role of ecotones in ecological systems, special emphasis was placed on those ecotones occurring at the terrestrial/aquatic interface (such as gallery forests, wetlands, oxbow lakes, littoral lake zones, and areas of substantial groundwater–surface water exchange).

The four key discussion questions for the MAB research program are (Naiman et al. 1988b, Naiman et al. 1989):
1.   What is the importance of ecotones in maintaining local, regional and global biodiversity?
2.   Which functions do ecotones have, and to what degree do ecotones exert filter effects?
3.   Is there any influence of ecotones on the stability of adjacent patches?
4.   How have humans contributed to ecotone maintenance and restoration in the past?

In May 1988, an international workshop on "Land/inland-water ecotones: strategies for research and management, a first step for an integrated international programme" was held in Sopron, Hungary. Sponsored by the Hungarian Academy of Science, the International Institute of Applied Systems Analysis, Unesco/MAB, and Unesco/IHP, 75 participants from 26 countries took part in the week-long meeting. The Sopron workshop was concerned with the interface between research and management, and more specifically was charged with achieving the following goals:

-- to produce a synthesis of scientific information on land/inland water ecotones;
--to identify gaps in information and understanding, in respect to both scientific hypotheses and the needs of management;
--to explore directions for future collaborative research and action; and
--to develop a research prospectus with testable hypotheses.

The first task of the Sopron workshop was to prepare a state-of-knowledge synthesis based on the presentation and discussion of 12 commissioned reviews. These papers dealt with: ecological importance of ecotones; internal and external processes influencing the origin and maintenance of ecotones; applicability of models; landscape processes; lotic, lentic, wetland and groundwater ecotones; role of ecotones in aquatic landscape management; social and economic implications of ecotones. These presentations have been edited and published as a volume in Unesco's Man and the Biosphere Book Series (Naiman and Décamps 1990).

Several preliminary hypotheses from the Sopron meeting have been assigned to Unesco's MAB Scientific Advisory Committee for further discussion and clarification. Included among the hypotheses are (Holland and Décamps 1989):

a.   Community composition and structure in the ecotone are more sensitive to changes in hydrology than to changes in nutrient availability;
b.   The influence of the ecotone on adjacent systems is proportional to the length and shape of the contact zones;
c.   Most ecotones can be managed for long-term sustainable yields; and,
d.   Ecotones can be managed in ways which maintain their ecological integrity and economically viable biological diversity.

The Sopron workshop identified two major requirements for future development of an 'ecotone perspective:' an improved knowledge base and new tools and techniques. Needed improvements in the knowledge base include better understanding of ecotonal relationships between surface waters and groundwaters and greater comprehension of ecotones as natural systems that can speed the recovery process of damaged systems. The major tools and techniques needed include spatial and temporal statistical methods for the quantitative evaluation of patch and boundary dynamics (Naiman and Décamps 1990). In addition, the Sopron workshop called for sound management of land/inland-water ecotones, giving consideration to all factors affecting these ecotones and the drainage basins of which they are a part (Holland et al. 1990).

In September 1987, SCOPE's Scientific Advisory Committee on Ecotones established the scientific direction and organizational structure of the incipient SCOPE project on ecotones. The project was designed to focus on two key topics: the movements of energy, materials, and information across landscape boundaries; and patterns of biodiversity associated with ecotones. Three general questions were posed (Hansen et al. 1988):
1.   How do ecological system boundaries influence biotic diversity and the flows of energy, information, and materials?
2.   How will biodiversity and ecological flows associated with ecotones respond to environmental changes, especially in global climate, sea level, land use, and atmospheric trace gases?
3.   How should ecotones be managed within a changing environment?

Three subsequent workshops were planned on the general characteristics of ecotones, the responses of ecotones to global change, and the management of ecotones.

The first was held December 1988, and was entitled, "Influence of ecotones on biodiversity and ecological flows." This workshop included plenary papers that reviewed current knowledge on the structure, function, and classification of ecotones and patterns of diversity, ecological flows associated with ecotones, and quantitative methods for studying landscape boundaries (di Castri and Hansen *in press*). The resurgence of interest in ecotones has come at a technologically opportune time; the evolution of new equipment and methods has greatly increased the ability of scientists to quantitatively study ecotones (Johnston et al. in press).

Workshop participants provided recommendations on the contributions ecotones research can make to IGBP and discussed implications of ecotone research for landscape management (Hansen et al. 1988, di Castri and Hansen *in press*). For example, Risser (*in press*) noted that edges or ecotones are created when forests are harvested, but the cumulative ecological results of these cutting procedures depend upon harvest timing and spatial patterns.

CHARGE FOR 1989 SYMPOSIUM

Given the recent international interest in the ecotone concept, Ecological Society of America (ESA) members involved in the planning meetings perceived a need to share results of SCOPE and MAB discussions with our North American colleagues. In August 1989, ESA, with financial support from United States MAB, sponsored a symposium on "The role of landscape boundaries in the management and restoration of changing environments" at the ESA annual meetings at the University of Toronto in Canada. This symposium was the first opportunity for a large segment of the North American scientific community to discuss a scientific program on landscape boundaries. Since the theme for the 1989 ESA meeting was "Global Change," the symposium was particularly timely.

The symposium speakers were asked to:

1.   Examine environmental processes occurring in boundaries between landscape units;
2.   Relate this examination to development of IGBP;
3.   Produce recommendations to the U.S. National Committee for the MAB program to help guide future research.

In response to this three-pronged charge, speakers were asked to consider the following topics in their presentations:

a.   Fundamental ecological characteristics of landscape boundaries;
b.   Human impact on the functioning of

landscape boundaries;

c.   Potential responses of landscape boundaries to global environmental change;

d.   Simulation of the scale dependent effects of landscape boundaries on species persistence and dispersal;

e.   Climatic constraints and issues of scale controlling regional biomes; and,

f.   Restoration of human-impacted ecotones.

The ESA symposium provided a public forum at which the latest research evidence, theory, and methodology were reviewed, debated, and discussed. Speakers discussed the characteristics of ecotones, how ecotones respond to environmental change, and the ways in which ecotones can be most appropriately managed. Recommendations from this symposium volume will serve as a comprehensive planning tool for future MAB research.

ACKNOWLEDGEMENTS

The authors thank D. Correll, Y.L. Grossman, M.N. Jensen, C. Ledsky, R.J. Naiman, and an anonymous reviewer for comments on various drafts of this manuscript. Special thanks go to R.E. Soles, Executive Director, U.S. Man and the Biosphere Program for implementing the project; T. Fuad, Research Assistant, ESA Public Affairs Office, for reviewing all bibliographic citations; and to A.H. Thomas, Administrative Assistant, ESA Public Affairs Office, for preparing the manuscript. Financial support from the United States Man and the Biosphere Program is gratefully acknowledged.

LITERATURE CITED

Brundtland, G.H. 1987. Our Common Future. World Commission on Environment and Development. Oxford University Press, New York, New York, USA.

Clements, F.E. 1905. Research methods in ecology. University Publishing Company, Lincoln, Nebraska, USA.

Committee on Earth Sciences. 1989. Our changing planet: the FY 1990 research plan: executive summary. United States Global Change Research Program. Office of Science and Technology Policy. Washington, D.C., USA.

DeFries, R.S., and T.F. Malone, editors. 1989. Global change and our common future: papers from a forum. National Academy Press, Washington, D.C., USA.

di Castri, F., A.J. Hansen, and M.M. Holland, editors. 1988. A new look at ecotones: emerging international projects on landscape boundaries. Biology International, Special Issue **17**:1-163.

di Castri, F., and A.J. Hansen, editors. *In press.* Landscape boundaries: consequences for biotic diversity and ecological flows. SCOPE book series. Springer-Verlag, New York, New York, USA.

Hansen, A.J., F. di Castri, and P.G. Risser. 1988. A new SCOPE project. Ecotones in a changing environment: the theory and management of landscape boundaries. Biology International, Special Issue **17**:137-163.

Holland, M.M. 1988. SCOPE/MAB technical consultations on landscape boundaries: report of a SCOPE/MAB workshop on ecotones. Biology International, Special Issue **17**:47-106.

Holland, M.M., and A.J. Hansen. 1988. Meeting Reviews: ecotones. Bulletin of the Ecological Society of America **69**(1):54-56.

Holland, M.M., and H. Décamps. 1989. A new international programme: research and management of landscape boundaries. Pages 102-103 *in* J.C. LeFeuvre, editor. Proceedings of the Third International Wetlands Conference. Conservation and development: the sustainable use of wetland resources. Museum of Natural History, Paris, France.

Holland, M.M., D.F. Whigham, and B. Gopal. 1990. The characteristics of wetland ecotones. Pages 171-198 in R.J. Naiman and H. Décamps, editors. The ecology and management of aquatic-terrestrial ecotones. Man and the Biosphere Series. The Parthenon Publishing Group, Carnforth, United Kingdom.

Johnston, C.A., J. Pastor, and G. Pinay. In press. Quantitative methods for studying landscape boundaries. In F. di Castri and A. Hansen, editors. Landscape boundaries: consequences for biotic and ecological flows. SCOPE book series. Springer-Verlag, New York, New York, USA.

Naiman, R.J., H. Décamps, J. Pastor, and C.A. Johnston. 1988a. The potential importance of boundaries to fluvial ecosystems. Journal of the North American Benthological Society 7:289-306.

Naiman, R.J., M.M. Holland, H. Décamps, and P.G. Risser. 1988b. A new UNESCO programme: research and management of land/inland water ecotones. Biology International, Special Issue 17:107-136.

Naiman, R.J., H. Décamps, and F. Fournier, editors. 1989. The role of land/inland water ecotones in landscape management and restoration: a proposal for collaborative research. Man and the Biosphere Digest 4. UNESCO, Paris, France.

Naiman, R.J., and H. Décamps, editors. 1990. The ecology and management of aquatic-terrestrial ecotones. Man and the Biosphere Series. The Parthenon Publishing Group, Carnforth, United Kingdom.

Odum, E.P. 1971. Fundamentals of ecology, third edition. W.B. Saunders Company, Philadelphia, Pennsylvania, USA.

Risser, P.G.. 1985. Spatial and temporal variability of biospheric and geospheric processes: research needed to determine interactions with global environmental change. Report of a workshop sponsored by SCOPE, INTECOL, and ICSU. 18 October-1 November 1985. St. Petersburg, Florida, USA.

Risser, P.G. In press. Implications of ecotone research for landscape management. In F. di Castri and A. Hansen, editors. Landscape boundaries: consequences for biotic diversity and ecological flows. SCOPE book series, Springer-Verlag, New York, New York, USA.

FUNDAMENTAL ECOLOGICAL CHARACTERISTICS OF LANDSCAPE BOUNDARIES

JAMES R. GOSZ. Biology Department, University of New Mexico, Albuquerque, New Mexico 87131, USA.

Abstract.   Identification of <u>fundamental</u> characteristics of landscape boundaries is difficult because of a lack of quantitative information on functional characteristics of boundaries. It is compounded by scale-dependent characteristics that often are not appreciated by the investigations. This review attempts to summarize general characteristics of boundaries and identify the research difficulties that must be addressed before fundamental characteristics can be identified. Research difficulties are grouped under the following topics: boundary detection, scale, scale-dependent measurement, and extrapolation and predictability. Because of the current interest in using boundaries to evaluate climate change, the characteristics of biome boundaries also are addressed and the Long Term Ecological Research Program at Sevilleta presented.

Key words: landscape boundary, scale-dependent characteristics, Long Term Ecological Research Program, Sevilleta.

INTRODUCTION

A SCOPE/MAB working group recently defined an ecotone as a: *Zone of transition between adjacent ecological systems, having a set of characteristics uniquely defined by space and time scales and by the strength of the interactions between adjacent ecological systems* (Holland 1988). At present there are no compelling arguments for differentiating between the terms "ecotone", "landscape boundary" and "transition zone" (Hansen et al. 1988) and this paper uses them interchangeably. Holland (1988) continues the definition with a discussion of the term "ecological system" as including commonly described hierarchical entities such as demes, populations, communities, ecosystems, landscapes and biomes. Thus, ecotones can be described as transitional zones or boundaries between ecosystems, or between biomes, etc. Regardless of what terminology is employed, it is important to keep in mind that boundaries are identifiable and meaningful only relative to specific questions and specific points of reference. What appears as an ecotone at one spatial scale may be seen as a collection of patches at a finer scale.

The definition used in the opening paragraph was recognized by the authors as being general; however, they agreed to such a general definition as a launching point for development of the theoretical base necessary for future discussions of the ecotone concept. Holland (1988) identifies the recent widespread recognition and call for the study of ecotones are based on the following four suppositions: 1) the number of putative characteristics of ecotones that are significant in understanding ecological systems in general; 2) the assumption that ecotones are highly susceptible, and are thus good early indicators of changes; 3) the potential significance of ecotones for prudently managing the biosphere; and 4) the recognized relative paucity of data from ecotones. The comparative lack of data from ecotones is compounded by a dearth of techniques for explicitly measuring the dynamic processes characteristic of these transition zones.

General Boundary Characteristics

Even casual observation reveals that most landscapes are composed of various components. A typical rural landscape might include several agricultural croplands, pastures, woodlands, streams, farmsteads, and roads. Thus, the landscape is heterogeneous, that is, consists of dissimilar or diverse components or elements. Recognition of those features also infers the recognition of edges or transitions between the features. This trivial sounding statement has very significant implications; recognition is in the eye of the beholder. The human bias will play a strong role in how boundaries are recognized, measured, and interpreted. Different individual scientists may recognize different boundaries and, clearly, different organisms recognize different boundaries.

The traditional view of an ecotone is an intersection between plant communities where there is a relatively abrupt change in vegetation structure (di Castri et al. 1988). One of the early references described an ecotone as a tension zone where principal species from adjacent communities meet their limits (Clements 1905). In terrestrial systems, boundaries are generally recognized operationally by spatial discontinuities in features of the soil and/or vegetation; abrupt transitions between woodlands and grasslands or between riparian and desert vegetation are typical examples. The outer band of an ecosystem or patch, referred to as the edge (or edge effect, see Forman 1986), contains an environment significantly different from the interior area, and hence also typically differs in biomass, soil characteristics, and species composition and abundance. Unfortunately, many ecological studies over recent decades avoided the complexities of transitions and worked in the centers of relatively homogeneous areas (patches). Many studies also ignored the interactions between such patches. Recent work has recognized the heterogeneity of ecological systems and the typical mosaic of components differing in structure and dynamics (Pickett and White 1985). Collectively, the dynamics of the components of the mosaic strongly influence the characteristics of the entire system (Shugart 1984, Hansen et al. 1988). Although recent attention focusing on boundaries identifies the value of such an approach; there is little developed theory.

Ecotones can possess specific abiotic and biotic characteristics, such as physical and chemical attributes, biotic properties, and energy and material flow processes. Although these characteristics are used in defining ecosystems and biomes, the unique conditions of ecotones are most significant in their influence on the interactions with adjacent ecological systems. For example, the internal structure of a forest edge adjacent to an open area is dominated by a shrub and small tree zone called a mantel, that controls the penetration of outside weather into the forest. Just outside the mantel is a perennial herb layer or saum (Forman 1986). These zones differ in their degree of development and density depending upon the location of the disturbance line relative to the outermost tree branches. Additional zones appear to exist within certain edges, and marked variations in edge structure along the border of a forest are evident. These internal edge zones and variations, analogous to a cellular membrane, doubtless play key roles in the movement of species between ecosystems (Wiens et al. 1985). Thus the movement of animals is concentrated at certain spots, probably due to both edge structure and species composition. Similarly, wind, solar radiation and moisture are funneled into a forest only at certain points along a forest edge (Forman 1986). Some ecological flows across ecotones may be unidirectional, moving only from one patch to another, while others are bidirectional. The strength of these interactions, which may vary over time and space scales, may be driven by the contrast between adjacent ecological units (i.e., the distinctions among the characteristics and the dynamics of the relevant processes within the adjacent ecological systems and the ecotone itself; Holland 1988).

In terms of system dynamics, boundaries are locations where the rates or magnitudes of ecological transfers (e.g. energy flow, nutrient exchange) change abruptly in relation to those within the patches. Identifying the locations based on ecosystem processes may be difficult because they demonstrate considerable variability in the transition zone (Naiman et al. 1988). Chaos theory, a mathematical concept predicting unpredictability, has been suggested to model the variability of transitions (Naiman et al. 1988, Gosz and Sharpe 1989). The premise of chaos theory is that chaotic, or seemingly unpredictable, behavior is in reality a special manifestation of an underlying structure. Models of chaos are usually in the form of difference equations which exhibit regular behavior at some values of one or more variables and irregular behavior at others. Naiman et al. (1988) found that such equations predict that boundaries can behave in ways that are not simple averages of adjacent resource patches, implying that the spatial patterning of resources and interactions between patches within the boundary are as important determinants of boundary behavior as are the adjacent patches.

To understand landscape-level patterns and processes, we must know the details of "boundary dynamics" - what determines why a boundary is located where it is; how boundaries influence ecological processes within patches and over the larger landscape; how boundaries affect the exchanges or redistribution of materials, energy, and organisms between landscape elements; and how these transfers can, in turn, act to change the location and nature of boundaries (Wiens et al. 1985). This is a distinctly different focus from the older concepts of "ecotone" and "edge effect," which primarily have considered how transitions between community types influence diversity and species distributions.

Wiens et al. (1985) originally drew the analogy of boundaries between elements in a landscape to membranes in organismal or physical systems. Like membranes, boundaries vary in their permeability or resistance to flows. This variation is a consequence of characteristics of the boundary itself (e.g., its thickness, the degree to which the separated patches differ) and of the responses of different materials, organisms, or abiotic factors to the boundary. Boundaries may thus be quite impermeable to transfers of some materials but may be permeable or "leaky" to other fluxes. Clearly, the boundaries separating elements in a landscape will have important influences on system properties both within homogeneous patches and between the components of a landscape.

The configuration and dynamics of boundaries in a landscape may diverge from the patterns anticipated on the basis of physical-edaphic gradients due to the actions of various "vectors", which are physical or biotic forces that actively move materials or energy in the system (Forman 1981). Vectors are important because they may impose nonrandom, directional fluxes of materials on the general background of resource concentration gradients and passive diffusion; they can actively move materials against a gradient. Vector movements can be influenced by boundary characteristics, for example by deflecting or blocking vector movements. The degree to which a boundary deflects the movements of vectors may be expressed as the boundary "permeability" and it may determine the degree to which the landscape element or patch can be viewed as a closed or an open system (Wiens et al. 1985). Abiotic vectors such as wind and water are strongly affected by features of the physical structure of the boundary (e.g. soil texture, topography). Wind movements across a boundary between a patch with low, sparse vegetation and one with greater vertical structuring and coverage, as occurs at some grassland-shrubland or grassland-tree boundaries, will be altered by the increased turbulence and decreased velocity produced by the more structured plant canopy.

Two generalities regarding influences of abiotic vectors on landscape boundary dynamics were derived from studies of the spatial and temporal heterogeneity of beaver-created patch bodies and their associated fluxes (Johnston and Naiman 1987). First, the permeability of a lateral boundary to abiotic vectors is a function of the kinetic energy of the vector; both wind and water have higher transporting capacities at higher velocities. Second, within-patch retention of particulate matter transferred by abiotic vectors across lateral boundaries is maximized by a decrease in kinetic energy. For waterborne transfers, this would occur where the patch boundary coincides with a decrease in slope, such as the boundary between an upland stream and a beaver impoundment. For wind-borne transfers, a decrease in kinetic energy would occur where there is change in vertical structure such as a row of trees.

Among biotic vectors, characteristics of the animals themselves contribute to boundary permeability. Given a relationship between home range size and body size, larger species within a taxonomic unit will probably find a given boundary more permeable than smaller species (Wiens et al. 1985). Physiological capabilities also strongly influence the abilities of organisms to occupy the different environments found in adjacent patch types. Susceptibility to heat or water stress, for example, can prevent some vectors from crossing a boundary between a patch containing adequate shelter and one with greater exposure. Differences in predation pressure between landscape elements may restrict movements between patches by predation-susceptible forms, while having relatively little influence on species with well-developed predator defenses. The likelihood that a boundary will be crossed by a vector is also related to the density of animals within a patch. While the probability that an individual will encounter (and perhaps cross) a boundary is a function of its movement patterns and position in the patch, increases in the density of the population within a patch produce a dramatic increase in the collective probability of boundary encounter. There is likely to be a clear density-dependence to between patch transfers in landscapes. The effects of the between-patch interactions and transfers produced by vectors will depend on the specific properties of each boundary situation and the particular vectors and resource distributions that characterize

the adjacent patches. The clearest effect may be the contrast between effects that lead to an overall stabilization of patch characteristics and an enhancement of patch boundaries and those that tend to destabilize patch features and erode patch boundaries. Thus, vectors for which a given patch boundary is relatively impermeable will have their movements deflected back into the patch by the boundary, leading to a localization of their activities within the patch. Establishment of central places (e.g. dens, colonies, nest sites) within the patch will accentuate this effect. Because vector activities are increasingly internalized within the patch, the boundary itself is likely to become sharper and more clearly defined. As a consequence, many system processes may become increasing self-contained within the patch, as both outputs to and inputs from other landscape elements may be reduced (Wiens et al. 1985).

Destabilizing influences tend to disrupt internal patch processes and boundary integrity by producing a new imbalance between inputs and outputs. One clear destabilizing effect is the spread of disturbances from a patch into the landscape surroundings. Since disturbances are less likely to spread over a spatially heterogeneous area than through a homogeneous region (but see Risser 1987), it follows that within a patchy landscape, disturbance spread should be inversely related to the sharpness of the boundary between patches.

Many examples illustrate that landscape heterogeneity may play a key role in the spread of disturbance. In general, greater heterogeneity inhibits disturbance spread. The spread of disturbance decreases as boundary-crossing frequency increases. This occurs with reference to animal movement (Forman and Godron 1986) as well as to disturbance spread in a landscape. Implicit here is that most edges (though not all) act as barriers or filters to the spread of disturbance, due to the special structure or combination of species present in an edge. Landscape heterogeneity, as measured by the types and relative abundance of ecosystems present, can be expected to affect the spread of disturbance. However, landscape configuration, that is, the spatial juxtaposition of patches, corridors, and matrices, is a more sensitive and precise controller of disturbance spread because it determines boundary-crossing frequency (Forman 1987).

Biome Boundaries

Relation to Climate:

The current interest in using boundaries to detect climatic change has stimulated much discussion. This was suggested early on in a study of plant communities (Griggs 1937) and has been developed further by paleobotanists using pollen and tree-ring analyses to identify patterns of climatological change in vegetation (Delcourt and Delcourt 1987). While natural ecotones or boundaries may be sensitive indicators of future climate change because they may reflect the climatic constraints on species (National Academy of Sciences 1988), human-caused boundaries may not play a useful role in this assessment other than indirectly through changes in human use of landscapes as the environment changes. The relationship between land use change and climate or "weather" will be a critical component of landscape ecology in many areas and humans may magnify the effect of climate change in predictable ways. For example, additional land in the United States was put into soybean production and timber clearing increased in Brazil for the production of soy as a result of the anchovy fishery decrease after the 1982-83 El Nino event (Barber 1988). Identifying "natural" boundaries that are sensitive to climatic constraints may be difficult in some areas because of undocumented, historical, human uses of the landscape. In any case, using boundaries to answer questions of climatic change must involve separation of "natural" from "human-caused" ecotones and "sensitive" from "unsensitive" natural ecotones.

Investigations of the problems of global change require that analyses and predictions be developed for very broad scales: regions, continents and the globe (National Academy of Science 1988). At the scale of regions, the appropriate boundaries to study are likely to be those of biomes or dominant life-forms. There are important features associated with these broad-scale boundaries that differ from fine-scale ecotones.

One approach to the study of biome boundaries is evaluating broad-scale environmental gradients that occur over a region and identifying boundaries that occur in the relatively steep portions of those gradients (Sala et al. 1988, Gosz and Sharpe 1989). This approach uses the steep gradients associated with biome boundaries to establish: 1) gradient relationships with distance; 2) whether spatial variability is dependent or independent of scale; 3) how gradient steepness influences system properties; and 4) if responses are integrated across the region (Gosz and Sharpe 1989). These ecological changes may be distributed over many kilometers. The structural features of fine-scale ecotones (hundreds of meters) are likely determined by site specific characteristics such as soil discontinuities, lake edge, and even fire. Climate appears as a constant across such small distances. Broad-scale boundaries or transition zones between biomes are more likely to be a result of large scale climatic features, such as gradients of temperature and moisture. On certain portions of the gradient, physiological thresholds will occur causing relatively abrupt changes in life-form (e.g., forest to prairie transition). The broad-scale change in climate accounts for more of the general variation in vegetation across such a transition zone than fine-scale features of the habitat. At these broad-scales, the pattern is one of relatively homogeneous biomes separated by transitions of varying abruptness. The physical structure of biotic communities at the broad-scale is primarily determined by climate with a secondary influence from soils.

Neilson et al. (*in press*) and Neilson (this volume) analyzed spatial patterns of seasonal weather associated with different climatic regions. Climate exists in quite large, homogeneous regions that can be divided latitudinally into general circulation cells, frequently referred to as Hadley cells, with the transitions demarcated by jetstreams. All of the major biotic regions and their boundaries appear to be controlled by major climatic divisions. For example, the shift from eastern deciduous forest to tall grass prairie corresponds to a shift in rainfall pattern from east to west. The tall grass prairie is climatically differentiated from the short grass prairie by a large drop in both summer and winter rainfall. Neilson (1987*a*) and Neilson et al. (*in press*) describe how bands of rainfall traverse regions in the spring and fall which produce the grassland and desert biomes, respectively. This is an example of how broad-scale mechanistic processes produce patterns at the same scale.

Climate regionalization is the major constraint producing a biome type. Within the biome will be regional patterns of biotic variation that are mediated by variation in substrate (e.g., topographic and edaphic) along regional gradients. Generally, there will be more variation in communities near a regional boundary as compared to those within a core region (Neilson and Wullstein 1983, Neilson 1987*b*). Where the regional climate is near optimal for a given set of species (i.e., within the core region of a biome), minor differences between microhabitats will be within an organism's range of tolerance. A given species can range over a variety of slopes, aspects, and elevations resulting in large concentrations of the species. Where the regional climate is marginal, many microhabitats are outside the species' range of tolerance resulting in its distribution being constrained to fewer microhabitats of smaller size. When the size of a homogeneous patch is large relative to the individual (i.e., within the biome's core region), it can be occupied by many

different individuals (of the same or different species) which have overlapping limits of tolerance to the resource gradient. In this case, competition among individuals is likely to be diffuse (Neilson et al. *in press*). In boundary areas, where usable microhabitats (patches) may be small, only one or a few individuals might occupy any given patch, increasing the potential for inter- and intraspecific competition. Thus, the change in grain structure (i.e., patch size) along regional climatic gradients could modulate the kind and intensity of competition among organisms. Physical environments that are extremely heterogeneous in biome boundary regions (i.e., complex topography and soils) accentuate these patterns.

The patterns described above may move over the landscape in response to shifts in climate. Grover and Musick (1990) documented Chihuahuan shrub expansion patterns in New Mexico, presumably caused by grazing pressure and climate variation or change. The boundary of the desert shrub zone moved northward over the recent 50 year interval, and there was an increase in the proportion and size of habitats occupied by the desert shrub community in southern New Mexico (Fig. 1). As the shrub community moved northward, previous boundary areas became more like the core area of this community, and more microhabitats had conditions that fell within the tolerances of the desert shrub and outside the tolerances of the semi-arid grassland community, the previous dominant community type.

On a longer timeframe, Delcourt and Delcourt (*in press*) analyzed a 20,000-year time-series of paleoecological data with changing ecoclines across eastern North America and drew the following conclusions: 1) Although an ecotone may appear fixed in location through time, it yet may change in other attributes (strength, breadth) as community composition changes across it; 2) The number, position, and strength of ecotones between major vegetation types can all change given a strong enough environmental forcing function; 3) Different

13

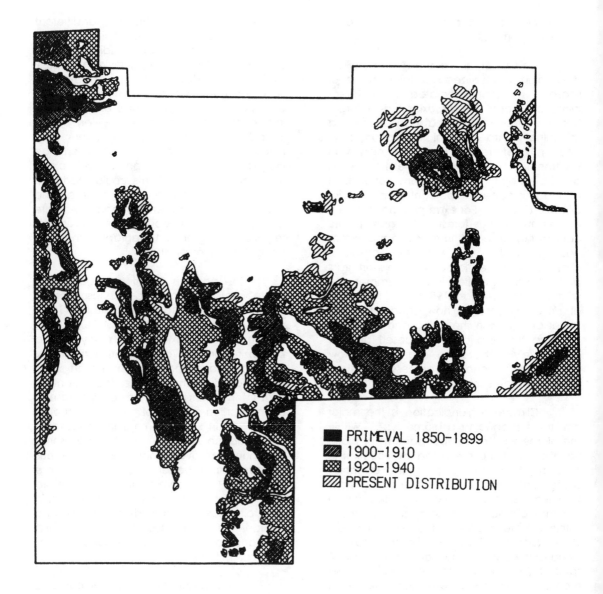

PRIMEVAL 1850–1899
1900–1910
1920–1940
PRESENT DISTRIBUTION

Figure 1. Southwestern New Mexico distribution of desert shrub communities (creosote bush) from 1850 to present. The biome boundary for this community has moved northward and the proportion of microhabitats allowing desert shrub establishment has increased (data from D.J. McCraw, in Grover and Musick, 1990).

ecotones may exhibit different sensitivities to the same environmental forcing function, for example depending upon proximity to migration fronts for populations of invasive species; 4) Ecotones may be ephemeral if an environmental change causes a restructuring or disassembly of communities that were responsible for definition of the ecotones; 5) Ecotone dynamics on one boundary of a community may be different than along another boundary of the community; 6) At the macro-scale domain in space and time, there is no paleoecological evidence that ecotones confer stability to the communities they bound.

The temporal variability of weather, irrespective of any long-term change in climate, should influence local diversity patterns, especially in boundary zones. Weather regimes that promote the establishment of one component of the community (e.g., shrub life-form) may persist for one to several decades and then shift into another pattern that promotes a different life-form (e.g., grasses). Neither life-form may suffer local extinction if it has life-history characteristics (e.g., perennial structures, seed storage) that allows persistence during the unfavorable period, at least in certain microhabitats. After some period, the weather may shift back to re-establish the former vegetation. Thus, the local diversity can be enhanced by variations in weather in biome boundary areas. Likewise, attempts to relate today's diversity to today's weather may be inappropriate (Neilson 1986, Neilson et al. *in press*). The consistent movement of the desert shrub in New Mexico along with the "filling in" of more and more habitats of former boundary areas indicates that either a general climatic and/or land use trend (grazing) has occurred.

Ecosystem Properties

At some point along a climatic gradient, life-form changes and a new biome is encountered. At the boundary, the change in life-form is often the result of reaching the distributional limits of one to several generalist species or genera that are capable of extending the full range of the biome (e.g., semi-desert shrub *Larrea*). This results in a steep gradient in plant structural properties and a change in ecosystem process rates at the edge of such distributions (Gosz and Sharpe 1989). Changing weather or climate conditions that result in movement of a new life-form across a landscape or altered spatial patterning in such biome transition zones markedly affect ecosystem properties at a range of scales. Often a change in life-form is a change in plant structure or a change in lignin quantity and distribution. Such community structural differences markedly affect decomposition and nutrient cycling characteristics (Gosz 1981, Berg and McClaugherty 1987) as well as properties such as ecosystem biomass, nutrient storage, resistance and resilience, and microclimate (Odum 1971).

Primary productivity is strongly related to the environmental forcing functions of temperature and moisture plus other system characteristics resulting from the environmental functions (e.g., soil organic matter). Patterns of primary productivity are expected to parallel environmental gradients over regions and biomes (Sala et al. 1988). Biomass, however, is a result of net primary productivity, life history strategy, and disturbance history. Biomass may approximate annual net primary productivity for some biomes (e.g., annual grasslands) or be orders of magnitude larger for long-lived woody perennials, for which biomass is a net accumulation of many years of net productivity. The relatively sharp transition in life-form (i.e., life-history strategy) at a biome boundary infers that boundary zones between biomes have very steep gradients in biomass and in the ecosystem properties that are affected by these structural differences. The movement of boundaries with their associated changes in community structure may cause marked, nonlinear responses of ecosystem properties to changes in weather regimes or climate that are much larger than accompanying changes in net primary productivity. Such boundary zones may be sites of attenuation or amplification of ecosystem processes; properties that magnify the influence of subtle changes in the environmental forcing functions into measurable ecosystem responses (Gosz and Sharpe 1989).

Sevilleta – A Case Study

The Sevilleta National Wildlife Refuge in central New Mexico is part of the National Science Foundation's network of long term ecological research (LTER) areas in the United States. It is a very large site (approximately 100,000 ha) and is located in a transition zone between three major biomes: Great Plains grassland, Great Basin shrub-steppe, and Chihuahuan Desert. A fourth biome (conifer woodland; Mogollon flora) occurs at higher elevations on the Sevilleta and forms transitions with the other three biomes at lower elevations. Vegetation characteristic of all four biomes occur on the Sevilleta as well as various levels of overlap between the species representing the different biomes. The result is many novel assemblages of species and life-forms. The site has been described as especially valuable for quantifying: 1) gradient relationships with distance for different scales; 2) how spatial variability depends on scale; 3) how steep environmental gradients influence system properties; 4) the amplification or attentuation of system properties in transition zones between biomes; and 5) the use of transition zones as sensitive measures of climatic change (Gosz, in press). The system is ideal for studying complexity because it includes: 1) strong spatial patterning caused by steep gradients in major abiotic constraints; 2) fluctuating driving variables (e.g. El Nino/La Nina episodes) which produce differential responses of biota throughout the site; 3) very high plant and animal diversity (resulting from having heterogeneous topography and being a transition zone of four biomes; and 4) direct or inverse relationships (plus threshold responses) between the biota and environmental factors.

Topography, geology, soils, and hydrology, interacting with climate dynamics,

provide the spatial and temporal template on which the interactions occur. At least 880 plant species and varieties are found within the boundaries of the Sevilleta (Manthey 1977) and fifty-four species terminate their geographic distribution within the study site. The area represents not only the boundaries of species distributions, but also boundaries for various life-forms (e.g., desert shrub, conifer tree, perennial grass) and physiologies (e.g., C_3 perennial grasses). At least 75 species of mammals, 207 land birds, 59 reptiles, and 16 amphibians are present; a substantial proportion of these species have a geographic distribution boundary near or in the Sevilleta (Hubbard 1970, Findley et al. 1975, H. Snell unpublished). Reptiles provide the most dramatic example, as 47 of the 59 species at Sevilleta end their distributions in the vicinity of the Sevilleta (33 of these are northern limits of desert species).

Precipitation is a very significant forcing function on the Sevilleta. There are a number of precipitation gradients in the area contributing to vegetation boundary formation. Relatively stable gradients and boundaries exist where topography strongly controls air mass dynamics (Neilson 1987). Other boundaries are dynamic and under the influence of spatially and temporally varying periodicities in climatic factors (e.g., 40 yr, 20 yr, 4-5 yr, Gosz, in press).

The biome boundaries that characterize Sevilleta demonstrate complex and nonlinear dynamics. Gosz and Sharpe (1989) used a catastrophe model as a conceptual tool for studying these transition zones. The qualitative descriptions of the four properties of a catastrophe model are (Jones 1977):

1. Bimodality – the system at the low precipitation levels typical of the Sevilleta tends to be either in one biome type or another. Gradients between biome types are steep (i.e., small spatial distance relative to the scale of each biome), and intermediate conditions tend not to occur. In this case the transition occurs over a 10 to 20 km gradient;

2. Discontinuity - as the controlling factor (e.g., soil texture, soil moisture holding capacity) varies toward either extreme, vegetation type varies markedly. An example is the jump that occurs from one system property to another with very small differences in soil moisture holding capacity;

3. Divergence - a small difference in annual precipitation could result in very different community structure over the range of low to high soil moisture holding capacity; and,

4. Hysteresis - the response made as the control factors move in one direction is different from the response made as the control factors move in the opposite direction.

A precipitation threshold for primary productivity is an important factor in the nonlinear dynamics at Sevilleta. The inverse texture hypothesis of Noy-Meir (1973) is applicable where annual precipitation fluctuates around 370 mm (Sala et al. 1988). Above this precipitation threshold, productivity increases with soil moisture holding capacity (SMHC) characteristic of fine textured soils. If precipitation is below 370 mm, productivity decreases with SMHC because fine textured soils resist water penetration, and evaporation occurs before plants make effective use of the resource. Coarse textured soils allow deeper penetration of scant moisture, thereby reducing evaporation. At higher precipitation quantities, there is a gradual relationship of increased productivity with increased SMHC.

One example of hysteresis for the Sevilleta relates to episodes of drought (e.g., 1950's) that resulted in high mortality in conifer woodland species at their distributional limits (i.e., primarily at the low elevations and/or on south aspects). Those areas now have standing dead juniper, much of the grass understory is gone or greatly reduced, and desert shrub (*Larrea tridentata*) is present. It is probable that the drought conditions caused the conifer-grass mortality and allowed the desert shrubs that now dominate those sites to become

established. A simple return to higher annual precipitation has not allowed the conifer woodland species to reinvade. The species will require low temperatures plus abundant winter precipitation, perhaps following several years of summer drought.

These conditions plus the different life-form and species responses to environmental forcing functions on the Sevilleta may allow unique tests of catastrophe models for a biome boundary area. These models may relate current conditions on the Sevilleta and the dynamic response of its communities to fluctuations in climate, which are important aspects of the LTER research program for this site.

Research Difficulties in Boundaries

The paucity of information as well as the generality of the definition used for boundaries makes the identification of Fundamental Ecological Characteristics of Landscape Boundaries a difficult task. A different approach to identifying fundamental characteristics is a discussion of **fundamental difficulties** in dealing with landscape boundaries. Issues of scale, scale-dependent results, assumptions of methods, and technique-dependent results plague all fields of ecology. They are no less problematic in the ecological dynamics of landscape boundaries. In addition, they are augmented by issues of edge definition, edge detection, and how scale-dependent processes are influenced by edge identification. The issue is further complicated because human-caused ecotones (e.g., agricultural fields or forest edges resulting from ownership boundaries, topographic features, historical events) may be lumped with natural ecotones where abiotic constraints restrict species distributions. The remainder of this paper will focus on these **fundamental difficulties** as they will identify the research needed to establish the fundamental characteristics of landscape boundaries. The primary issues are: 1) Boundary detection; 2) Scale; 3) Scale-dependent measurement; and 4) Extrapolation and predictability.

Boundary Detection

Turner et al. (1990) discuss results from two, 2700 m transects on a desert watershed with an elevational gradient, extending from the edge of a rocky mountain slope to a low-basin ephemeral (playa) lake. Boundaries were characterized for annual and perennial vegetation (from Ludwig and Cornelius 1987, Wierenga et al. 1987), ants, reptiles, birds, and mammals (Fig. 2a-d). These ground sampling data demonstrate the different properties of boundaries proceeding from abiotic controlled zonation upwards through several trophic levels. A moving window technique was used which averages a certain number of sample points along the transect and determines values for each half of the window. Large differences in values for the two measurements at a window location (the peaks in Fig. 2) reflect an edge or boundary (Turner et al. 1990). Window sizes of 6 and 12 sample points provided similar results for vegetation. Six discontinuities were found which defined seven vegetation zones. These vegetation discontinuities were strongly coincident with soil discontinuities determined using the same technique (Wierenga et al. 1987).

The figures also contrast peak or edge locations along the transect for different consumer groups. For example, at a given window length, many more edges or boundaries appear in the organization of the ant species than in mammals, which are groups usually classified at the same trophic level, but which exhibit different mobility. There appear to be 4 distinct edges (= 5 zones) for mammals, corresponding closely in location with some (but notably, not all) of the perennial plant zones identified by Wierenga et al. (1987). Perennial plants (predominately grasses) exhibit two distinct zones at the upper end of the transect, but these are not reflected in the edge pattern for mammals. The heterogeneity demonstrated (at this scale) for ants does not correspond to any lower level patterns, nor to any apparently meaningful structure.

Similarly, edge patterns differ between lizards and birds (lizards being strictly predatory, while birds include both granivorous and predatory species). Using this scale for interpretation, bird species observed across this watershed appear to exhibit two community boundaries (= 3 zones) that include combinations of 2-3 perennial plant zones. Thus, the distributional pattern for birds (a mixed-consumer level group) does indeed appear more spatially integrative than the other consumer groups.

In another study, Johnston et al. (*in press*) reported on the use of the moving split-window technique for identifying functional boundaries not visibly obvious related to below ground nutrient cycling processes in a beaver pond in northern Minnesota, USA. In this study performed by Pinay and Naiman, window sizes of 6 and 10 samples were used to analyze soil water data collected at two different depths on three sample dates from a 54 m transect. The analysis revealed that some of the boundaries, such as the aerobic boundary between the pond and its *Carex* border, were very stable and coincided well with visible physical and vegetative changes. However, other boundaries were more ephemeral in location and did not coincide with visible changes in the environment. Furthermore, boundaries detected in surface soil did not always coincide with those deeper in the soil profile. This study of the dynamics of aerobic/anaerobic ecotones showed that an ecotonal area may actually consist of numerous ecotones, both structural and functional, that may be temporally asynchronous, and may or may not be visibly detectable.

These valuable data sets display the complexity that can exist among organisms, boundaries of organism activities, and potentially, ecosystem processes in a given landscape. Questions as fundamental as *What is a boundary? Where is a boundary?* have no simple answer other than *It depends!* Typical images in our minds or in textbooks are of sharp contrasts (e.g., forest

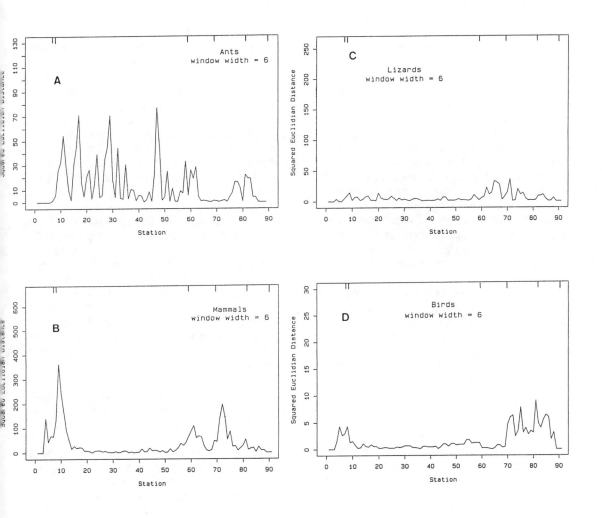

Figure 2. Hierarchy of zonation on a desert transect (from Turner et al. 1990). Results of moving-window edge analysis, using raw (abundance) data Squared Euclidean Distance (d), width window width 6 for (A) ants, (B) mammals, (C) lizards, and (D) birds. Data are from a 2700 m transect with sampling stations at 30 m intervals with stations numbered consecutively beginning with the lower elevation (North) end of transect. Ticks at top of each plot show location of zone boundaries of perennial vegetation as identified by Ludwig and Cornelius (1987).

edge, agricultural field boundary). Other organisms may view these contrasts differently which will affect the interactions between adjacent ecological systems.

Scale

A number of studies imply that natural scales exist in ecological phenomena, and that selection of a sampling unit size appropriate to the phenomena will yield the most accurate information. Measurement of ecological processes at an appropriate scale is analogous to viewing a television screen (Carlile et al. 1989). Viewing the screen closely enough to see individual dots, or measuring on a small scale, may reduce the ability to discern the image formed collectively by the dots. Conversely, viewing the screen from too great a distance or measuring on too large a scale may result in a blurred image. It is necessary to determine the most appropriate measurement scale for different processes or physical properties (e.g., different scales of measurement may be necessary to study the various processes influencing a given population of a species).

Scale may be especially important in boundary dynamics since it is a part of the definition of the boundary or ecotone. Landscape ecology, and therefore landscape boundaries, cannot escape dealing with spatial analysis, spatial scale and scale-change effects. A landscape may appear to be heterogeneous at one scale but quite homogeneous at another scale, making spatial scale inherent in definitions and landscape heterogeneity and diversity (Meentemeyer and Box 1987, Gosz and Sharpe 1989). For example, Risser (1987) identified a number of "principles" for landscapes:

1. The relationship between spatial pattern and ecological processes is not restricted to a single or particular spatial or temporal scale.

2. Ecological processes vary in their effects or importance at different spatial and temporal scales. Thus, biogeographic processes may be relatively unimportant in determining local patterns but may have major effects on regional patterns. For example, processes leading to population decline may produce extinction at a local scale, but may be manifest only as spatial redistributions or alterations in age structure at broader geographic levels (Risser 1987, Wiens 1989).

3. Different species and groups of organisms (e.g., plants, herbivores, predators, parasites) operate at different spatial scales; therefore, investigations undertaken at a given scale may treat such components with unequal resolution (see Fig. 2a-d). Each species views the landscape differently, and what appears as a homogeneous patch to one species (including humans) may comprise a very heterogeneous patchy environment to another (Risser 1987, Wiens 1989).

4. Scales of landscape elements are determined by the specific objectives of the investigation or the pertinent management question. If a study or management issue focuses at a specified scale, processes and patterns occurring at much finer scales are not always perceived because of filtering or averaging effects, whereas those occurring at much broader scales may be overlooked simply because the focus is within a smaller landscape unit.

One challenge of landscape ecology is contending with the large array of spatial and temporal scales of ecological processes and disturbance regimes: there is no fundamental level of investigation (O'Neill et al. 1986). For example, in considering the mosaic of grassland landscapes, there are processes ranging from square millimeters to hundreds of kilometers and from time scales of minutes to millenia (Delcourt et al. 1983, Risser 1987, Delcourt and Delcourt *in press*). The transect study reported by Turner et al. (1990) demonstrates the complexity of multiple-scale sampling. Figure 3 is a comparison of different window sizes (different scales) on the detection of mammal and ant boundaries along the

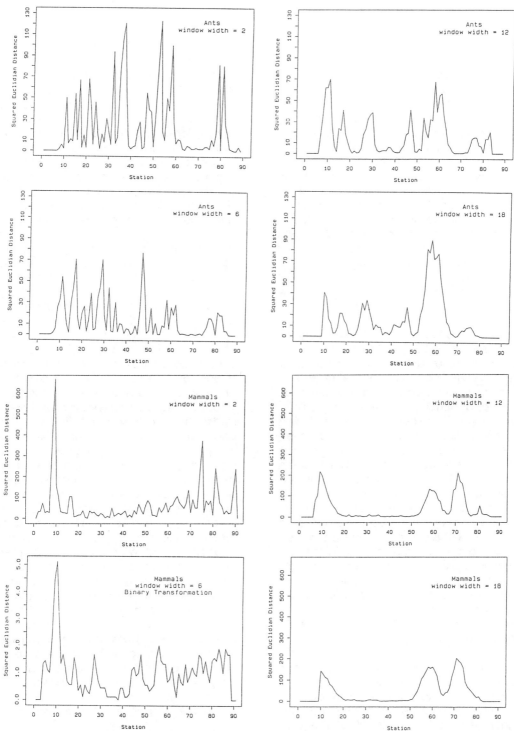

Figure 3. Ecological edges for ants and mammals at windows (w) = 2, 6, 12, 18 (unpublished Figures, courtesy of M. Conley and W. Conley). Resolution of the moving window edge analysis is maximum at w = 2 and minimum at w = N. Data for mammals are relatively robust to changing w, while the ant data are more dynamic under changing w. The question that requires attention involves the choice of w for any given group that "best" represents ecological arrangement and organization.

21

transects described earlier. Mammals exhibit essentially the same boundary locations for window widths 2-18. The relative sizes of the peaks marking boundaries change dramatically; for window widths 2-12, the major shift in communities is identified at the lower end of the transect (the playa boundary), while for window width 18, the major discontinuities are boundaries along the upper part of the transect. In contrast, boundary analyses for ants identify different locations for each window width, 2-18 with the patterns for window sizes 12 and 18 providing equally useful interpretations. The relative magnitude of discontinuities shifts among the edges identified for each window width.

In another study Rusek (*in press*) examined assemblages of soil flora and fauna at several scales along three transects. Total diversity increased or decreased in what was determined to be the center of an ecotone, depending in part, on the scale of the ecotone. In the case where diversity was depressed (i.e., grassland-forest ecotone), there were actually several zones across the ecotone, separated by smaller-scale ecotones. The soil organisms exhibited considerable pattern and species turnover at each of these zones in the ecotone. His study suggested an important objective for dealing with ecotones; sampling at multiple scales. Both of the above examples identify this as a powerful approach for identifying the scale of different patterns or measurement techniques and for identifying those processes and patterns that recur at all scales versus those that are unique to particular scales.

Merriam and Wegner (*in press*) related the importance of structural differences in the habitat to population dynamics. Populations in isolated patches cannot be considered without reference to neighboring patches that can contribute propagules. Although a population of some species in a single patch might go locally extinct, recolonization can maintain the "metapopulation" (Gill 1978). Population dynamics are therefore determined by the initial conditions within patches and the movement of organisms among patches. These kinds of population dynamics differ from those of large, continuously distributed populations. At an ecotone, there is often a gradient from a single, large connected patch to ever smaller and less connected patches. There may be thresholds along this gradient where demographic processes suddenly shift, and a species becomes endangered or where allelic frequencies suddenly change. Such thresholds may depend on both organism scale and patch scale. Some species might view an ecotonal landscape as a patch, for example, whereas others might not. Differences in effective dispersal distance could mediate these different responses.

Ecotonal phenomena are evident over a broad spectrum of space and time scales and result from a hierarchy of environmental constraints. Ecologists need to adopt a multiscale perspective and perform studies at several scales or in which grain and extent are systematically varied independently of one another to provide resolution of domains, patterns and their determinants, and of interrelationships among scales. An important focus must be on linkages between domains of scale. Meentemeyer and Box (1987) recently called for the development of a "science of scale" in ecology and Wiens (1989) recommends that we consider scaling issues as a primary focus of research efforts. Our ability to arrange scales in hierarchies does not mean that we understand how to translate pattern-process relationships across the nonlinear spaces between domains of scale. We only recognize such linkages when we identify how different hierarchical levels constrain one another (Wiens 1989).

Scale-dependent Measurement

In addition to scale-dependent properties across ecotones, using different measurement techniques or different variables in a given area may cause additional technique-dependent results. The different window sizes of Figure 3 provide one example of how results differ with different scales but the measurement technique was not changed. When different techniques are used, each operating at a different scale, additional types of measurement-dependent results can occur because of different assumptions for the different techniques. Figure 4 demonstrates how the boundary patterns for mammals in Figure 2 change when presence/absence values or frequency values are used. In Figure 2, Squared Euclidean Distance is used and potential maximum distance values are a function of number of variables (species) sampled. Reporting presence/absence values, thereby removing the influence of widely varying maximum abundances, causes distinct boundaries at the upper end of the transect to disappear into a large zone of smaller scale heterogeneity. Using frequency values (scaling each species' abundance at a position to total abundance across all species occurring at that position), some of the edges identified in Figure 2 are apparent, but other, new peaks appear in the area previously identified as homogeneous.

Meentemeyer and Box (1987) discuss the results of Thornthwaite and Mather (1955) that demonstrate how different sized evaporation devices, especially small containers, can produce exceptionally biased results. Very small evaporating surfaces better represent the humidity of the atmosphere than the true evapo-transpiration rate of a vegetated area or evaporation of a lake. Under high humidity conditions, no scale-dependency occurs; however, as humidity decreases, larger discrepancies occur for small-scale devices. When environmental processes (e.g., evapotranspiration, denitrification, primary production) are quantified for boundary or transition situations, both apparatus and measurement must be evaluated carefully for scale-dependent features.

The general approach to quantifying characteristics of boundaries, or even landscapes, must be reevaluated. Traditional multivariate statistical methods have to be used with caution (Meentemeyer and Box 1987). According to Haggett et al. (1977) location data are generally spatially autocorrelated, non-stationary, non-normal, irregularly spaced and discontinuous. Spatial data probably violate all rules for parametric statistical analyses.

Extrapolations and Predictability

A principal challenge for all fields of ecology is extrapolation and prediction from study sites to larger scales. It is generally assumed that studies at broader scales average the local (fine-scale) heterogeneity and are more predictable (O'Neill et al. 1986). One of the principal difficulties in extrapolating from fine-scale studies, even without scale-dependent effects, is having sufficient replication at the fine scale to correctly estimate the average condition that would correspond to a broad-scale analysis. At fine scales, many factors (and their interactions) can influence relationships, causing site-specific results. At a broader scale, many of the site-specific results average out, leaving broader scale patterns that reflect broad-scale constraints (e.g., climate, Neilson et al. *in press*).

Wiens (1989) discusses factors affecting the ability to predict ecological phenomena. At broader scales, the time scale of important processes also increases because processes functioning at those scales operate at slower rates, time lags increase, and indirect effects become increasingly important (see also Delcourt et al. 1983, Clark 1985). The hypothetical relationships between spatial and temporal scales of variation (Fig. 5) suggest that any predictions about the dynamics of spatially broad-scale systems that do not have the appropriate temporal scale are pseudopredictions (Wiens 1989). The predictions may seem robust, but because they were made on a fine time scale relative to the dynamics of the system, the mechanistic linkages and dynamics will not be seen. For example, a long-lived conifer tree does not appear to change on a daily time scale. If we study processes in that tree that do change on a daily temporal scale

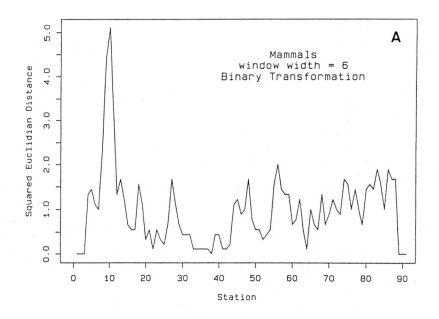

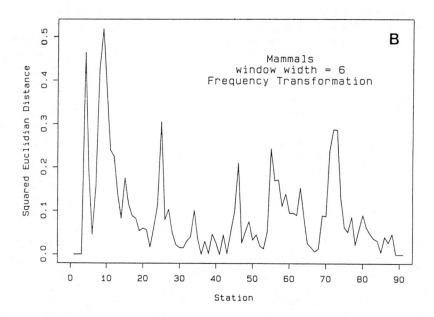

Figure 4. Edge detection using different transformations (unpublished Figures, courtesy of M. Conley and W. Conley). Results of moving-window edge analysis at window width 6 using data for mammals: (a) converted from abundance to presence-absence values; and (b) converted to frequency values by dividing each species's abundance at a station by total abundance of all species at a station. Compare these to the mammals window 6 raw data (abundance) plot in Figure 2. Note that the frequency data above demonstrates changing shape of the mammal assemblies along the transect.

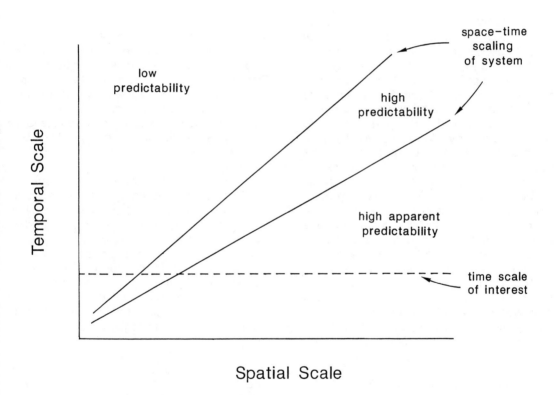

Figure 5. As the spatial scaling of a system increases, so also does its temporal scaling, although these space–time relationships differ for different systems. Studies conducted over a long time at fine spatial scales have low predictive capacity. Investigations located near to the space–time scaling functions have high predictive power. Short-term studies conducted at broad spatial scales generally have high apparent predictability (pseudopredictability) because the natural dynamics of the system are so much longer than the period of study (from Wiens 1989).

(e.g., stomatal behavior, phloem dynamics relating to carbon gain), we must study at fine spatial scales and will find tremendously detailed behavior and variability. Generalizations about how the entire organism will change are more likely to emerge at broader scales, probably measured by techniques appropriate to those scales (e.g., biomass increment of different tissues on an annual or decadal time scale). Since an objective is to extrapolate to broader scales, long-term investigations are necessary to reveal the dynamics of the system.

The following simple example suggests how and which landscape boundaries can detect phenomena such as climate change. Climate change is a broad-scale feature, and appropriately, has a long temporal behavior pattern. In Figure 5 such a feature would be located in the upper right corner of the graph. Only a long-term and broad-scale study is likely to correctly assess climate change. Although many researchers suggest that local ecotones can be used to detect climate change, such fine-scale individual studies can only reflect weather (finer-scale and shorter temporal scale events). Extrapolating local ecotone dynamics to the scale of climate change results in pseudoprediction. However, if a network of ecotones over a large area were studied for a long period, it is possible that the variability due to site-specific factors and local weather would average over the long term and reveal a climatic pattern. Meteorologists have long used this approach.

RECOMMENDATIONS/NEEDS

The previous sections dealing with the difficulties associated with understanding boundaries has identified a number of recommendations and needs. The following list of specific items augments that discussion.

*We must develop multiple scale analyses of boundaries. Nested watersheds provide analyses of scale-dependent behavior for stream discharge, nutrient output, species distributions, etc., and a similar approach using nested patterns of patch aggregation should provide scale relationships across boundaries. The innovation involves developing a true measure for processes of the larger scale aggregations.

*Spatial patterns in landscapes violate rules for parametric statistics. We must acknowledge that landscape units and boundaries have nonnormal, spatially autocorrelated, nonstationary, discontinuous, and irregularly spaced parameters. We need additional mathematical and statistical techniques.

*A classification system should be developed for boundaries/ecotones. Human-caused boundaries are likely to react differently to environmental change than other boundaries. Boundary sensitivity is a critical factor; however, not all boundaries are equally sensitive to different changes. We must identify the underlying principles that cause sensitivity.

*Social/economic factors often lead to human-caused boundary formation and change. Ecologists are not prone to relate the human dimension to ecological studies; however, more and more landscapes will be controlled because of social and economic pressures, and that influence will be as important as physical or environmental influences. The major influence of climate change may be the altered human manipulation of landscapes.

*Managing landscapes and boundaries means reducing some types of variation deemed unfavorable, adding constraints that control ecological processes, etc. We must research how we can build human views and management desires into both boundary dynamics and within patch dynamics. Conversely, an important objective is to develop principles for boundary dynamics that will influence human views and management.

*The recent focus on the ecology of boundaries may represent a danger in that we may be going from one extreme to another. In the same way that studying only homogeneous regions misrepresents landscapes, concentrating on boundaries can also misrepresent the landscape. Gradients are ever present and contrasting only the flattest and steepest (i.e., patch interior and edge) is insufficient. We must develop relationships for changes in structural and functional characteristics across all degrees of gradient change.

ACKNOWLEDGEMENTS

This manuscript was improved considerably by the reviews of D. Correll, M.M. Holland, M.N. Jensen, R.J. Naiman, and an anonymous reviewer. The support of the Ecological Society of America was instrumental in convening the symposium and leading to the discussions and interactions that led to the manuscript. This is contribution no. 8 to the Sevilleta LTER program.

LITERATURE CITED

Barber, R.T. 1988. Ocean basin ecosystems. Pages 171-194 in L.R. Pomeroy and J.J. Alberts, editors. Concepts of ecosystem ecology. Springer-Verlag, New York, New York, USA.

Berg, B., and C. McClaugherty. 1987. Nitrogen release from litter in relation to the disappearance of lignin. Biogeochemistry 4:219-224.

Carlile, D.W., J.R. Skalski, J.E. Batker, J.M. Thomas, and V.I. Cullinan. 1989. Determination of ecological scale. Landscape Ecology 2:203-213.

Clark, W.C. 1985. Scales of climate impacts. Climatic Change 7:5-27.

Clements, F.E. 1905. Research methods in ecology. University Publishing Company, Lincoln, Nebraska, USA.

Delcourt, P.A., and H.R. Delcourt. 1987. Long-term forest dynamics of the temperate zone. Springer-Verlag, New York, New York, USA.

Delcourt, H.R. , P.A. Delcourt, and T. Webb, III. 1983. Dynamic plant ecology: The spectrum of vegetation change in space and time. Quaternary Science Review 1:153-175.

Delcourt, P.A., and H.R. Delcourt. In press. Ecotone dynamics in space and time. In F. de Castri and A. Hansen, editors. Landscape boundaries: consequences for biotic diversity and ecological flows. SCOPE book series. Springer-Verlag, New York, New York, USA.

di Castri, F., A.J. Hansen, and M.M. Holland, editors. 1988. A new look at ecotones: emerging international projects on landscape boundaries. Biology International, Special Issue 17:1-163.

Findley, J.S., A.H. Harris, D.E. Wilson, and C. Jones. 1975. Mammals of New Mexico. University of New Mexico Press, Albuquerque, New Mexico, USA.

Forman, R.T.T. 1981. Interaction among landscape elements: a core of landscape ecology. Pages 35- 48 in S.P. Tjallingii and A.A. de Veer, editors. Perspectives in landscape ecology. Proceedings of the International Congress. Netherlands Society of Landscape Ecology, Netherlands.

Forman, R.T.T. 1986. Emerging directions in landscape ecology and applications in natural resource management. Pages 59-88 in R. Herrmann and T. Bostedt-Draig, editors. Proceedings of the Conference on science in the National Parks. Volume 1. Colorado State University, Fort Collins, Colorado, USA.

Forman, R.T.T. 1987. The ethics of isolation, the spread of disturbance, and landscape ecology. Pages 213-229 in M.G. Turner, editor. Landscape heterogeneity, and disturbance. Springer-Verlag, New York, New York, USA.

Forman, R.T.T., and M. Godron. 1986. Landscape ecology. Wiley & Sons, New York, New York, USA.

Gill, D.E. 1978. The metapopulation dynamics of the red-spotted newt, *Notophthalmus viridescens* (Rafinesque). Ecological Monographs **48**:145-166.

Gosz, J.R. 1981. Nitrogen cycling in coniferous ecosystems. Pages 405-426 in Clark, F.E. and T.R. Rosswall, editors. Terrestrial nitrogen cycles; processes, ecosystem strategies and management impacts. Ecological Bulletins, Volume 33, Stockholm, Sweden.

Gosz, J. *In press*. Ecological functions in a biotransition zone: translating local responses to broad scale dynamics. *In* F. de Castri and A. Hansen, editors. Landscape boundaries: consequences for biotic diversity and ecological flows. SCOPE volume. Springer-Verlag, New York, New York, USA.

Gosz, J. R., and P.J.H. Sharpe. 1989. Broad-scale concepts for interactions of climate, topography and biota at biome transitions. Landscape Ecology **3**:229-243.

Griggs, R.F. 1937. Timberlines as indicators of climatic trends. Science **85**:251-255.

Grover, H., and B. Musick. 1990. Shrubland encroachment in southern New Mexico, USA: an analysis of desertification processes in the American Southwest. Climate Change **16**:165-190.

Haggett, P., A.D. Cliff, and A. Frey. 1977. Locational analysis in human geography. Edition 2. Wiley & Sons, New York, New York, USA.

Hansen, A.J., F. de Castri, and R.J. Naiman. 1988. Ecotones: what and why? Biology International, Special Issue **17**:9-46.

Holland, M.M. 1988. SCOPE/MAB Technical consultations on landscape boundaries: report of a SCOPE/MAB workshop on ecotones. Biology International, Special Issue **17**:47-106.

Hubbard, J.P. 1970. Check-list of the birds of New Mexico. New Mexico Ornithology Society, Publication Number 3, Albuquerque, New Mexico, USA.

Johnston, C.A., and R.J. Naiman. 1987. Boundary dynamics at the aquatic-terrestrial interface: the influence of beaver and geomorphology. Landscape Ecology **1**:47-57.

Johnston, C.A., J. Pastor, and G. Pinay. *In press*. Quantitative methods for studying landscape boundaries. *In* F. di Castri and A. Hansen, editors. Landscape boundaries: consequences for biotic diversity and ecological flows. SCOPE book series. Springer-Verlag, New York, New York, USA.

Jones, D.D. 1977. Catastrophe theory applied to ecological systems. Simulation **29**:1-15.

Ludwig, J.A., and J.M. Cornelius. 1987. Locating discontinuities along ecological gradients. Ecology **68**:448-450.

Manthey, G.T. 1977. A floristic analysis of the Sevilleta Wildlife Refuge and the Ladron Mountains. Masters Thesis, University of New Mexico, Albuquerque, New Mexico, USA.

Meentemeyer, V., and E.O. Box. 1987. Scale effects in landscape studies. Pages 15-34 in M.G. Turner, editor. Landscape heterogeneity and disturbance. Springer-Verlag, New York, New York, USA.

Merriam, G., and J. Wegner. *In press*. Local extinctions, habitat fragmentation and ecotones. *In* F. de Castri and A. Hansen, editors. Landscape boundaries: consequences for biotic diversity and ecological flows. SCOPE book series, Springer-Verlag, New York, New York, USA.

Naiman, R.J., H. Décamps, J. Pastor, and C.A. Johnston. 1988. The potential importance of boundaries to fluvial ecosystems. Journal of the North American Benthological Society 7:289-306.

National Academy of Sciences. 1988. Toward an understanding of global change. National Academy Press, Washington, D.C., USA.

Neilson, R.P. 1987a. Biotic regionalization and climatic controls in North America. Vegetatio 70:135-147.

Neilson, R.P. 1987b. On the interface between current ecological studies and the paleobotany of pinyon-juniper woodlands. Pages 93-98 in R.L. Everett, editor. Proceedings of the Pinyon-Juniper Conference. United States Department of Agriculture, Forest Service General Technical Report INT-215. Ogden, Utah, USA.

Neilson, R., G. King, R. DeVelice, and J. Lenihan. In press. Regional landscape boundary patterns: the responses of vegetation to subcontinental air masses. In F. de Castri and A. Hansen, editors. Landscape boundaries: consequences for biotic diversity and ecological flows. SCOPE book series. Springer-Verlag, New York, New York, USA.

Neilson, R.P., and L.H. Wullstein. 1983. Biogeography of two southwest American oaks in relation to atmospheric dynamics. Journal of Biogeography 10:275-297.

Noy-Meir, I. 1973. Desert ecosystems: environment and producers. Annual Review of Ecology and Systematics 4:25-51.

Odum, E.P. 1971. Fundamentals of ecology. W.B. Saunders, Philadelphia, Pennsylvania, USA.

O'Neill, R.V., D.L. DeAngelis, J.B. Waide, and T.F.H. Allen. 1986. Hierarchical concept of ecosystems. Princeton University Press, Princeton, New Jersey, USA.

Pickett, S.T.A., and P.S. White, editors. 1985. The ecology of natural disturbance and patch dynamics. Academic Press, New York, New York, USA.

Risser, P.G. 1987. Landscape ecology: state of the art. Pages 1-14 in M.G. Turner, editor. Landscape heterogeneity and disturbance. Springer-Verlag, New York, New York, USA.

Rusek, J. In press. Distribution and dynamics of soil organisms across ecotones. In F. de Castri and A. Hansen, editors. Landscape boundaries: consequences for biotic diversity and ecological flows. SCOPE book series. Springer-Verlag, New York, New York, USA.

Sala, O.E., W.P. Parton, L.A. Joyce and W.K. Lauenroth. 1988. Primary production of the central grassland region of the United States. Ecology 69:40-45.

Shugart, H.H. 1984. A theory of forest dynamics: the ecological implications of forest succession models. Springer-Verlag, New York, New York, USA.

Thornthwaite, C.W., and J.R. Mather. 1955. The water balance. Publications in Climatology 8:1-104.

Turner, S.R., R.V. O'Neill, W. Conley, M. Conley, and H. Humphries. 1990. Pattern and scale: statistics for landscape ecology. Pages 17-49 in M.G. Turner and R.H. Gardner editors. Quantitative methods in landscape ecology: the analysis and interpretation of landscape heterogeneity. Springer-Verlag, New York, New York, USA.

Wiens, J.A. 1989. Essay review: spatial scaling in ecology. Functional Ecology 3:383-397.

Wiens, J.A., C.S. Crawford and J.R. Gosz. 1985. Boundary dynamics: a conceptual framework for studying landscape ecosystems. Oikos **45**:421–427.

Wierenga, P.J., J.M.H. Hendricks, M.H. Nash, J. Ludwig, and L.A. Daugherty. 1987. Variation of soil and vegetation withdistance along a transect in the Chihuahuan Desert. Journal of Arid Environments **13**:53–63.

CLIMATIC CONSTRAINTS AND ISSUES OF SCALE CONTROLLING REGIONAL BIOMES

RONALD P. NEILSON. Oregon State University, United States Environmental Protection Agency, Environmental Research Lab, Corvallis, Oregon 97333, USA.

Abstract. The prospect of climatic change threatens to cause large changes in regional biomes. These effects could be in the form of qualitative changes within biomes, as well as spatial changes in the boundaries of biomes. The boundaries, or ecotones, between biomes have been suggested as potentially sensitive areas to climatic change and therefore useful for monitoring change. Regional gradients of vegetation habitat size and variability are explored for their utility in detecting ecotone location and movement as driven by climatic change. Maximal habitat variability, as indicated by differential survivorship of plants, occurs at the ecotones or transitions between biomes. The principles developed for the analysis of abrupt changes in spatial habitat patterns (ecotones) will also be considered for the analysis and detection of potentially abrupt physiognomic changes through time (thresholds) over large regions. Extensive regional changes in habitat variability could occur rapidly, indicating an impending 'ecotone' in time (threshold) over much of the region. Rapid, regional changes could produce significant negative impacts on biological diversity. The two types of change, boundary shifts of regions and physiognomic shifts within regions, are potentially independent and may require different monitoring strategies to detect impending change.

Keywords: biodiversity, biogeography, biomes, climate change, ecotones, landscape variability.

INTRODUCTION

The prospect of rapid, human-caused global change has awakened a need for large-scale ecological science. The task of adapting to or mitigating the potential change could be enormous. Research needs should be prioritized, to identify sensitive regions and ecosystems and to define cost-effective management practices. Ecotones might be sensitive early warning systems of regional change, as well as being important systems in their own right (di Castri et al. 1988). They are, for example, being valued as landscape corridors for the migration of organisms, as barriers to (or facilitators of) flows of energy and matter across them, and as habitats of high biological diversity (di Castri et al. 1988).

The potential responses of biomes (the major vegetation types of Earth) to climatic change are the focus of this paper. The ecotones between forest and grassland, for example, could rapidly shift across the landscape. Such boundaries are readily observable from satellites and might, therefore, comprise an obvious monitoring tool for the impacts of climatic change. On the ground, however, direct observation of ecotones is less obvious. A theory will be developed in this paper relating regional climatic gradients to regional patterns of habitat variability and the utility of these patterns in (1) locating ecotones on the ground, (2) detecting change in these ecotones through space, and (3) the implications of these patterns of habitat variability to regional and local patterns of biodiversity of plants. The concepts developed from the analysis of spatial habitat variability will then be examined for their potential application to the variation of habitat patterns through time.

Changes in the ecological characteristics of entire biomes could accompany ecotone movement through space. Ecotone is a term usually reserved for abrupt spatial change in ecological communities. However, abrupt change through time over large regional extents appears quite possible without necessarily affecting the boundaries of the region (Neilson 1986, Neilson et al. 1989). Forested regions, for example, could convert to savanna. How might we anticipate potential physiognomic changes of biomes and would they happen slowly or rapidly? Can we use patterns of spatial habitat variability and the changes in that variability over time as an early warning of potential threshold changes in the structure and function of regional biomes? The term 'threshold' will be used interchangeably with 'ecotone in space,' but applies to the more general concept of abrupt change in either space or time.

The general approach will be to examine the scale of habitat patch size as 'perceived' by the organism. That is, a habitable patch for a tree could be restricted to the size of a north-facing slope, or it could encompass both north and south-facing slopes over a larger landscape. A change in the size of a suitable habitat patch will be referred to as a re-scaling of habitat size relative to a particular organism. Regional gradients in the scaling of physical habitat variability by organisms will be examined with regard to ecotones and regional climatic gradients.

The regional patterns of habitat scale (size) and variability (diversity), through space and time, will be explored as a potential indicator of proximity to ecotones or to regional ecosystem thresholds through time. Habitat diversity has long been recognized as a contributor to patterns of species diversity, different habitats fundamentally favoring different species (MacArthur 1972, Cody 1975). Habitat diversity should be assessed relative to the species of interest. For example, experimentally observed survivorship, growth and reproduction of a single species across nested gradients of physical habitats constitutes a bioassay and will reveal the 'realized' habitat variability as 'perceived' by the organism (Neilson and Wullstein 1983). If seedlings of a given species survive equally well on a range of qualitatively different microhabitats, the range of microhabitats is 'perceived' as homogeneous and the size of the habitat is large. The habitat will also be constrained to ecologically similar species. However, if the seedlings exhibit differential survivorship across a range of qualitatively different microhabitats, then some of the sites will be better suited to ecologically different species, thus promoting a regional increase in species diversity (Stebbins 1952, Axelrod 1967, Neilson and Wullstein 1983). The more habitats as 'perceived' by a single species, the richer should be the community composition.

Scaling of Habitat Patterns

Habitat pattern at the coarse scale is one of relatively homogeneous biomes separated by ecotones of varying abruptness. The distribution of biotic communities at the biome scale is primarily determined by climate with a secondary influence from soils (Whittaker 1975). That is, the climate determines if a region will support a forest or a grassland, while the soil influences local patterns of composition, health, and density. Therefore, physiognomic diversity should be predictable from environmental factors, while, prediction of species diversity will require the capability to predict physiognomy plus knowledge of biotic interactions and evolutionary and migratory history (Mooney 1977).

Diversity is usually considered to have two components, richness and evenness. Species richness is the number of different species, while evenness reflects their relative abundance (Pielou 1975). For simplicity, subsequent discussion will be restricted to species richness. Alpha diversity is defined as local diversity or species packing, typically that species richness measured in a small plot (Whittaker 1972, Cody 1975). Gamma diversity refers to the diversity of some arbitrarily large region, typically measured as total species richness among a number of small plots throughout a region. Beta diversity is the rate of change of species composition along a gradient and is usually measured by applying a similarity index to adjacent plots along a transect (Whittaker 1972, Cody 1975). At the continental or biome scale, species diversity patterns would be considered as gamma diversity and are often illustrated in contour maps of total species number over large geographic space (e.g. Simpson 1964, Cody 1975, Currie and Paquin 1987).

Delcourt and Delcourt (*in press*) demonstrated that beta diversity at the biome scale is relatively low in core areas and increases to a maximum at biome ecotones. They analyzed gradients of beta diversity across biomes, intersecting both core and ecotonal areas, and by constructing transects through time, examining periods of stasis and change. The temporal approach of examining stasis and change in species composition is analogous to the spatial analysis of core and ecotonal areas. In both cases the emphasis is placed on the rate of change of species composition along some gradient, either time or space. Beta diversity was shown to peak at ecotones in space and at rapid transition points, or thresholds of physiognomic change, in time (Delcourt and Delcourt *in press*).

I propose that the primary process controlling both the spatial and temporal gradients in beta diversity is the climatic constraint on the scaling of habitat space as 'perceived' by the organism. Regional climatic gradients produce regional gradients in habitat scale and diversity with minimum size and maximum diversity occurring at ecotones and at rapid climatic transitions in time (Delcourt and Delcourt *in press*). If the spatial and temporal gradients in beta diversity can be causally related to common gradients of habitat structure, we may be able to monitor the spatial and temporal patterns of habitat scale as an early warning signal of climatic change. One such relationship was demonstrated with a regional bioassay of microhabitat suitability along spatial and temporal climatic gradients (Neilson and Wullstein 1983).

Ecotones in Space - Regional Biomes

A plot (Fig. 1) of the different global biomes on a state-space of mean annual temperature and mean annual rainfall (Whittaker 1975) indicates that the biomes occupy broadly unique regions of the state-space. The Holdridge life-zone approach is a more complicated, but fundamentally similar, approach to that of Whittaker (Brown and Gibson 1983). For the sake of simplicity, I will restrict this discussion to the Whittaker approach.

There are two types of transitions in the diagram. First, there are large areas in the plot, particularly the semi-arid grassland region, where the biome could be one of a number of states from grassland to woodland depending on other conditions, such as seasonality or interannual variability of weather or human influence. The second type of transition (Fig. 1) is represented by the lines separating different biome types. These lines should also be considered as somewhat 'fuzzy,' in that constraints of climate seasonality and other factors, such as presence and intensity of grazing or fire, will produce physiognomic shifts in one direction or another (Whittaker 1975, Neilson 1986, Neilson 1987*a*). The seasonality and year-to-year variability of weather, as well as various disturbance regimes, should modulate the expressed state of the system within the constraints of annual temperature and rainfall (Allen and Starr 1982, Neilson 1986, O'Neill et al. 1986).

The Whittaker state-space diagram (Fig. 1) contains no spatial or temporal dimensions. The transition zones in the state-space diagram could represent rapid physiognomic transitions in space (ecotones) with a minor change in climate, or could imply a rapid physiognomic change in time over entire regions with a minor change in climate. The concepts presented in the diagram are equally valid in both dimensions of time or space. I will first discuss the spatial implications and then generalize to the temporal dimension. Also, by relating weather variability to the timing of plant life-history states and their seasonal physiological characteristics, mechanistic hypotheses relating climate to biome physiognomy can be constructed.

The spatial patterns of seasonal weather associated with the different biotic regions of

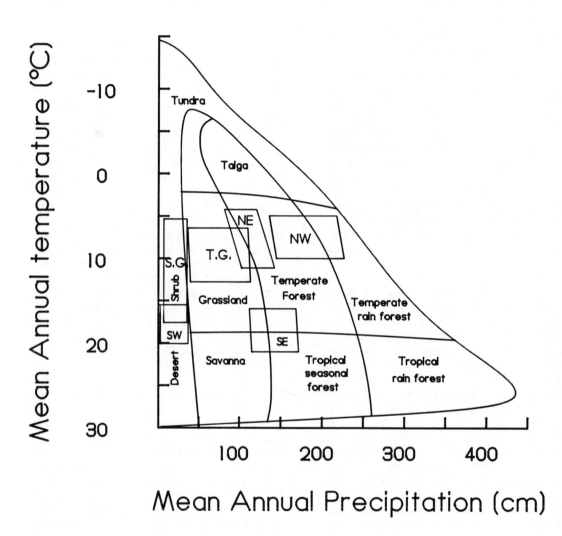

Figure 1. Plot of world biome types as a function of mean annual temperature and rainfall. The grassland area is broadly transitional and could be either grassland or some form of open woodland or scrub depending on other factors, such as climate, seasonality, soils or disturbance. The transitions between biomes as indicated by dividing lines are also approximate, depending on a similar set of other factors (after Whittaker 1975). Abbreviations for biome types are as follows: S.G. = short grass, T.G. = tall grass, S.W. = southwest deserts, N.E. = northeast hardwoods, S.E. = southeast pines and hardwoods, N.W. = northwest conifers.

the conterminous United States were analyzed by constructing a grid of line transects over the conterminous United States. Continental-scale transects were constructed through a network of weather stations (Quinlan et al. 1987) to examine the seasonal pattern and spatial extent of homogeneous climatic regions (see methods in Neilson and Wullstein 1983, Neilson et al. 1989). Regional and continental-scale runoff patterns were also examined in this manner using a United States Geological Survey stream gauge network (US WEST 1988). Average, monthly runoff from watersheds between 100-200 km^2 was normalized by the area of the basin and expressed in millimeters of depth. All of the major biotic regions and their ecotones are correlated with major climatic divisions or with the spatial extent of particular types of cold events, such as killing frosts (Neilson et al. 1989). The timing of rainfall events within each biome promotes the survival of specific life-history types and selects against other life-history types (Neilson 1986, Neilson 1987a).

Two transects delineate most of the potential climatic mechanisms that apparently control the major biomes across the United States (after Dice 1943, Küchler 1964, Neilson et al. 1989). For example, the eastern deciduous forest resides in a zone that receives high rainfall throughout the year (Figs. 2, 3). The transition or ecotone with the tall grass prairie occurs where the climate shifts to a summer rainfall peak and a winter rainfall minimum (Fig. 3). The tall grass prairie grades into the short grass prairie coincident with a large drop in summer and winter rainfall (primarily summer). The short grass prairie grades into the sage-scrub of the Great Basin as winter rainfall increases, while summer rainfall decreases even further. The sage-scrub gives way to the west coast forests as winter rainfall increases considerably at the Cascade Mountains. Although summers are generally dry on the west coast, it should be noted that the Cascades also delimit a large gradient in summer moisture, primarily in the form of humid, marine air (Tibbits 1979).

A transect across the southern tier of states produces a similar array of climatic and biotic zones (Figs. 2, 4). The southeast forest of pines and hardwoods is characterized by high rainfall throughout the year. The southeast forests give way to the southern plains at a regional climatic gradient where both winter and summer rainfall drop to very low values. The southern plains are characterized by spring and fall rains. The spring and fall rains almost cease at the eastern edge of the Sierra Madre Oriental – Rocky Mountain axis (ca. 105° W longitude, Fig. 2). West of this axis are the southwest deserts with spring and fall drought and mid-summer rains (Chihuahuan and Sonoran deserts). The interannual variability of winter rains in the southwest deserts appears to influence the relative dominance of grasslands and semi-desert shrublands in the region (Neilson 1986). The western region of the southwest deserts (the Mohave desert) receives little rain at any time of the year and extends to the coastal mountains. Winter rains again increase on the windward side of the mountains, but not to the amplitude of the Northwest.

The mechanism relating seasonal precipitation patterns to survival of different physiognomic types appears to reside in the regional water balance (Woodward 1987, Neilson et al. 1989, Stephenson 1990). Woody plants appear to rely upon a deep soil reservoir of water to balance the evapotranspiration demand during the growing season. This deep soil reservoir is re-charged each year primarily by winter precipitation. The transition between the eastern forested region and the Great Plains is clearly associated with the seasonality of runoff (Fig. 5, Neilson et al. 1989). The depth of runoff indicates the size of the deep soil reservoir of water. If winter rains are low, woody plants apparently cannot be supported and the abundance of spring and summer rains will determine the extent and biomass of grassland vegetation (Neilson et al. 1989).

A series of north-south transects, not shown, allow characterizations of the

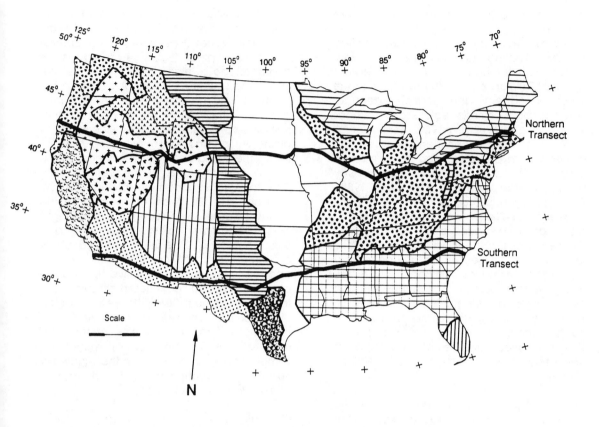

Biotic Regions

California Vegetation		Short Grass Prairie	
PNW Forests		Tall Grass Prairie	
Great Basin Sage-Steppe		South Texas Savanna	
Great Basin Desert Scrub		Northern Hardwoods	
Northern Rocky Mountain Forests		Eastern Deciduous Forests	
Southern Rocky Mountain Forests		Southeastern Forests	
Southwestern Deserts		Subtropical Forests	

Figure 2. Biomes of the United States (modified from Dice 1943, and Küchler 1964). The climatic transects are shown (Figs. 3, 4). The hydrologic transect (Fig. 5) closely follows the northern climatic transect.

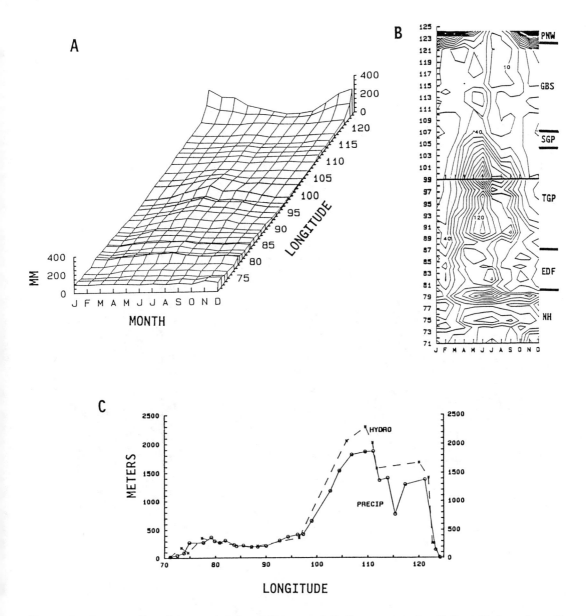

Figure 3. Seasonality of mean total monthly precipitation across the United States along the northern transect (Fig. 2): a) The vertical axis represents precipitation (mm/month), the horizontal axis represents months of the year, and the diagonal axis represents longitude. The locations of the weather stations used to construct the graph are indicated by horizontal lines. Total monthly precipitation data from 1941–1970 were used to calculate the mean values depicted here (Quinlan et al. 1987); b) Contour map of the three dimensional surface. The contour interval is 10 mm. Solid lines to the right of the contour plot indicate the positions of ecotones along the transect, as taken from Fig. 2. CV = California Vegetation, PNW = Pacific Northwest Forest, GBS = Great Basin Sagebrush–Steppe, GBD = Great Basin Desert Scrub, NRM = Northern Rocky Mountain Forests, SRM = Southern Rocky Mountain Forests, SWD = Southwestern Deserts, SGP = Short Grass Prairie, TGP = Tall Grass Prairie, STS = South Texas Savanna, NH = Northern Hardwoods, EDF = Eastern Deciduous Forests, SEF = Southeastern Forests; and, c) The elevations (m), progressing from East (low longitude) to West (high longitude), of the weather stations and of the stream gauging stations (Fig. 5).

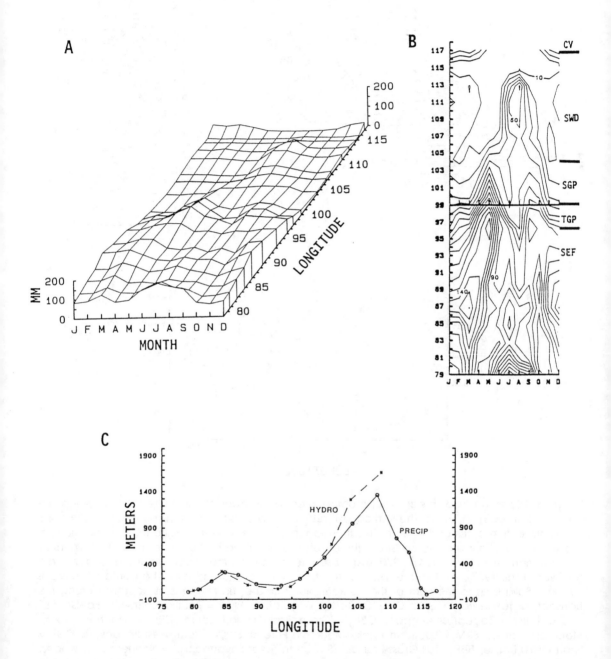

Figure 4. Seasonality of precipitation across the southern transect. Legends for (a), (b) and (c) are as in Fig. 3.

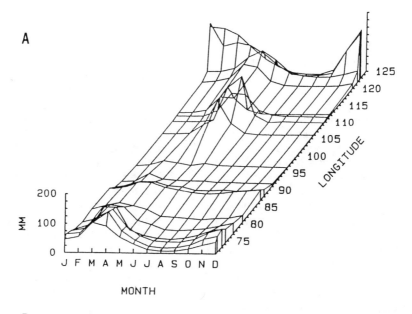

A

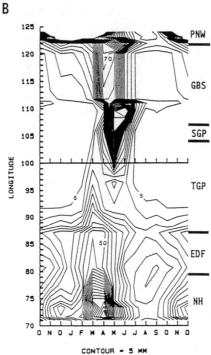

B

Figure 5. Seasonality of monthly runoff across the United States along the northern transect (Figs. 2,3): a) The vertical axis represents equivalent depth (mm), the horizontal axis represents months of the year, and the diagonal axis represents longitude, progressing from East (low longitude) to West (high longitude). The location of the gauging stations used to construct the graph are indicated by horizontal lines. At least 15 years of data were used to calculate the mean values (US West 1988); and, b) Contour map of the three dimensional surface. The contour interval is 5 mm. Solid lines to the right of the contour plot indicate the positions of ecotones along the transect, as taken from Fig. 2. See Fig. 3 for key to abbreviations. The elevations (m) of the gauging stations are shown in Fig. 3.

39

remaining biome-scale ecotones as functions of either rainfall seasonality shifts or physiological thresholds on thermal gradients, often seasonal frost phenomena (Neilson et al. 1989). A physiologial threshold is a temperature above or below which plants cannot survive. The north-south gradients exhibit more cold-temperature constraints than do the east-west gradients. The ecotones are usually defined by some thermal threshold in the physiological processes of plants along these continuous thermal gradients (Burke et al. 1976, Woodward 1987). Although the temperature gradients may not be steep, the ecotone may be sharply defined by virtue of the sharpness of the physiological threshold, such as the latitudinal extent of killing mid-winter frosts. The hardwood forest - boreal forest transition, for example, is controlled by the -40° isotherm, the anucleate freezing point of water (Burke et al. 1976). The ecotone between the eastern deciduous forest and the southeast pines and hardwoods may be controlled by the average extent of hard, winter frosts (Neilson et al. 1989). The ecotones along east-west transects in the conterminous Unites States are primarily related to the amplitude and seasonality of rainfall (Neilson et al. 1989). Thus, all major biome ecotones are associated with well-defined gradients in seasonal weather patterns, or physiological, threshold responses at critical points along thermal gradients.

Ecotones in Space - Local Habitat Patterns

Each of the broad biomes contain logically arrayed subdivisions (not shown) that reflect topographic, geologic, and land-use patterns. These subdivisions generally capture sub-biome variations in great soil groups, cropping systems, geologic substrate, and other large-scale features (Omernik 1987). These biome subdivisions are, in turn, composed of a mosaic of patches of different quality, arising from variations in soil or micro-topography or successional status (e.g., Whittaker and Levin 1977). Each of these patches can usually be further subdivided on some measurable

variability of habitat, and so on. Thus, spatial variability of habitat can usually be described as a hierarchy of different habitat types or qualities on a range of length scales (Allen and Starr 1982, O'Neill et al. 1986). Organisms will tend to 'perceive' a uniform habitat at one or another of these length scales (Neilson and Wullstein 1983). Furthermore, the scaling of habitat by organisms does not appear to occur randomly over a biome. Rather, there appear to be regional gradients in the scaling of habitat variability (Peet 1978, Watson 1980, Neilson and Wullstein 1983). Thus, local community associations appear to be under regional as well as local constraints (Neilson and Wullstein 1983, Ricklefs 1987).

Neilson and Wullstein (1983) described a theory of local habitat size and variability controlled by regional climatic gradients (see also Neilson 1987b). This theory, empirically developed, is quite similar to theories proposed by Stebbins (1952) and Axelrod (1967). In brief, if the regional climate is not overly stressful, organisms do not differentiate subtle changes in soils, topography or moisture availability. The concept of regional climatic controls on microhabitat suitability was tested in a transplant study of over 700 tree seedlings (Neilson and Wullstein 1983). Seedlings were planted across gradients of cover, aspect, elevation, and regional climate. Seasonal and year-to-year differences in seedling survivorship were also assessed. Physical measurements of soil moisture through a seasonal drought cycle demonstrated the same principles linking local site suitability and scale to regional climatic stress.

The range of available physical habitats for an organism will fall well within the tolerable physiological limits of the organism. Since minor habitat variability is not 'perceived' by the organism, the diversity of such habitats in the landscape will be comparatively small (Fig. 6). Minor physical differences between microhabitats will still be within the range of physiological tolerance and individuals of the species or of similar species will occupy essentially all

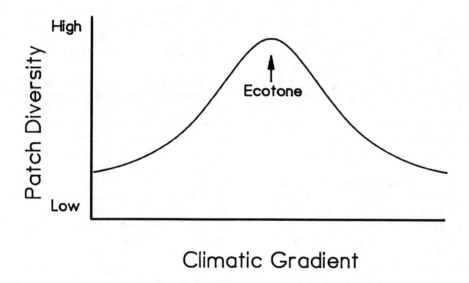

Figure 6. The hypothetical diversity of habitats along a transect stretching from one biome to another, across a major ecotone.

microhabitats in the landscape. However, if the regional climate is marginal, minor differences between microhabitats could place some of them outside the range of physiological tolerance, thus, the diversity of 'perceived' microhabitats in the landscape would be higher (Fig. 6). Species diversity increases under these conditions because those microhabitats that are outside the range of tolerance of the 'target' species will be within the range of tolerance of physiologically quite different species, often species from the neighboring biome. Thus, of the variation in geomorphic or edaphic habitats present in the environment, a comparatively larger proportion will be physiologically tolerable in core regions than near ecotones (Neilson and Wullstein 1986). Those contiguous patches of variable substrate that will support a given organism can be grouped into a larger 'homogeneous' habitat patch, accounting for the re-scaling of habitat size along regional gradients. An example will help to clarify these concepts.

Upper elevational boundaries of plant species in the central Rocky Mountains tend to be controlled by cold temperatures (Neilson and Wullstein 1983). This temperature constraint produces a latitudinal gradient in the upper bounds of species ranges (Fig. 7). The lower bounds of species in the central Rocky Mountains tend to be controlled by the amount of growing season rainfall. Since the moisture source is to the south (i.e., the Arizona Monsoon), there exists a latitudinal gradient in the lower bounds of species ranges, opposite in slope to the temperature gradient. The convergence of these two gradients in the northern part of the region constrains the tolerable patches to specific geomorphic facets and soils. That is, usable habitats are small in size and more constrained to particular topographic and soil conditions than further south. Seedling transplant data along local to regional gradients demonstrate the re-scaling of habitat size as a function of the regional climatic gradients (Neilson and Wullstein 1983).

Beta diversity, that is, the rate of change of species turnover along a gradient, measured in the southern part of the southern Rocky Mountains should be low compared to that measured further to the north, a result of the regional climatic control of the local suitability of microhabitats. The physical size of a habitable patch in the south is quite large and sweeps over minor variations in substrate and topography where the climate is nearly optimal. However, when the regional environment is stressful, due to drought or other factors, the organisms can survive in only small, select habitats. Thus, patch size will vary or re-scale relative to the organism along regional stress gradients (Neilson and Wullstein 1983, Neilson and Wullstein 1986). Core regions of biomes tend to be less stressful than ecotones, which are, almost by definition, at the limits of tolerance of the organisms on either side.

Is the model of regional control of local habitat size and variability a general model? The single requirement for the model developed in the central Rocky Mountains was that regional temperature and rainfall gradients were not parallel (Neilson and Wullstein 1983). This condition is ubiquitous for the biome ecotones of North America and likely throughout the world (Neilson et al. 1989). Although seasonal temperatures decrease smoothly toward the pole, seasonal rainfall amounts exhibit maxima and minima along any north-south transect. Therefore, at latitudinal ecotones, the temperature and rainfall gradients will not be parallel. Similarly, the east-west transects (Figs. 2, 3 and 4) are remarkable for the variations in rainfall, less so for temperature (data not shown). Again, the two stressors do not produce parallel gradients at any given ecotone. The coincidence of a steep gradient in either rainfall or temperature with known ecotones will be reflected by some form of "wedge" of tolerable environments (as in Fig. 7). Therefore, the pattern of increased beta diversity with proximity to ecotones should be a general phenomenon (Delcourt and Delcourt, *in press*).

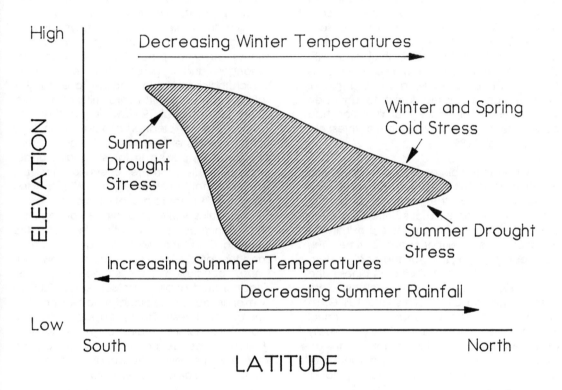

Figure 7. A conceptual model of the regional climatic constraints on the vertical distribution and composition of plant communities along a regional climatic gradient (after Neilson and Wullstein 1983, Neilson 1987b). As the gradients converge at the apex of the 'wedge,' the size of suitable habitat patches decreases, while their diversity increases (see Fig. 6).

Ecotones in Time – Physiognomic Thresholds

The Whittaker diagram of biome state-space (Fig. 1) is independent of space or time. Temperature or rainfall gradients in space appear to control ecotones between adjacent biomes. However, if the regional climate of a biome were near a physiognomic or ecologic state transition, the entire region could be sensitive to a rapid change of state with a minor change of climate. For example, the recent desertification of the southwest deserts rapidly shifted a large region from semi-desert grassland to desert shrubland with virtually no change in the biome boundaries (Neilson 1986). Similarly, a rapid, regional change from a closed forest to an open savanna would indicate that a region had passed through a physiognomic threshold (Neilson et al. 1989).

Although the biome diagram (Fig. 1) was constructed for all of the world's biomes, it is illustrative to plot the positions of several regions of the conterminous United States on the diagram to observe their proximity to transition zones in either space or time. This has been done (Fig. 1) for the northeastern hardwoods, the southeastern pines, the northwestern conifers, the southwestern deserts and the Short Grass and Tall Grass Prairies. Regional ranges of annual temperature and rainfall were visually extracted from national contour maps of climate (United States Department of Agriculture 1941) using the regional boundaries from Fig. 2 and plotted on the Whittaker diagram (Fig. 1). Note that the southwest and northwest regions plot diagonally from each other.

Several interesting observations emerge. The Tall Grass Prairie region is the only biome that falls well within one of Whittaker's zones, the grassland zone. Recall that this entire area of the diagram indicates a potential transition between grassland and woodland, depending on seasonality and other factors. All other regions plotted on the figure fall on or very near major transition zones. The northwest region is unique in that it is the only region that is not near a physiognomic transition, that is, it should always be forested. However, it could shift from temperate wet forest to temperate dry forest. Such a shift implies the potential for a large regional expansion of the interior pine forests into the mesic douglas fir forests, a fairly catastrophic ecological and economic change.

The entire eastern forested region could be transitional between temperate forest and temperate woodland, savannah, scrub or grassland, i.e., a major physiognomic change. Such a transition would be consistent with predictions of vegetative changes under global climate change, which indicate possible eastward expansion of grassland and northward shifts of all forest types (Neilson et al. 1989, Winjum and Neilson 1989). It is not clear what vegetation type might dominate the southeast region. Trees were not supported in forest models under various future climate scenarios for the southern half of the region (Winjum and Neilson 1989). There are, of course, numerous uncertainties with regard to these projections. These include uncertainties in both the General Circulation Models of global climate and the forest succession models. However, these conclusions are supported by an independent analysis (Neilson et al. 1989). If we examine the seasonal runoff patterns in the eastern forests (Fig. 5), it appears that the ubiquitous summer rains over the eastern United States are not sufficient to balance the extreme evapotranspiration over the region. Streams in the region are drawn down by regional evapotranspiration to a minimum flow in the summer when rainfall is still quite high (Neilson et al. 1989). The trees are apparently reliant upon a deep soil reservoir of water that is re-charged by winter rains. If either the winter rains decline or the summer transpiration increases due to increased temperatures, large regions of the country could experience summer soil moisture deficits. Such moisture deficits could trigger the kind of physiognomic transition implied by the Whittaker biome diagram (Fig. 1).

The Short Grass prairie and the southwest deserts receive similar amounts of rainfall and are both transitional between grassland and desert physiognomies (Fig. 1). Yet, the prairie appears to be more resilient as a grassland with few desert species, while the deserts are floristically more representative of true deserts. A regional and rapid transition from semi-desert grassland to desert shrubland in the southwest has already been observed and attributed to a combination of overgrazing and climatic variation (Neilson 1986). During the dustbowl of the thirties, the short grass prairie also clearly desertified with a cessation of spring rains (Neilson et al. 1989). However, unlike the southwest, the prairie region has returned to a grassland condition with the return of spring rains. It is interesting that the southwest desertification shows little signs of rebounding to grassland even with the amelioration of overgrazing. The lack of recovery may be due to the regional lack of the spring rains which appear to be important in favoring the physiognomy of grasslands (Neilson 1987a) and to significant changes in soil structure and fertility (Whitford et al. 1987). The differences between the two semi-arid regions could be due to their different patterns of rainfall seasonality, wet springs in the prairie and dry springs in the desert (Neilson 1987a, Neilson et al. 1989).

Thus, there could be thresholds of ecological change that pertain to entire regions. Spatial ecotones at the biome scale appear to be primarily controlled by rather sharp changes in seasonal climatic regimes or the spatial location of physiological temperature thresholds. In contrast, within biome, rapid temporal changes (thresholds) could be manifested as regional changes in the amplitudes of seasonal weather patterns (as distinct from changes in seasonality). For example, an increase in summer temperature and potential evapotranspiration could fully deplete a deep soil reservoir of water producing a threshold change in the physiognomy of a forested region with no change in regional rainfall patterns.

Ecotones that are constrained by winter or spring cold events could, in response to climatic change, tend to shift rapidly along shallow latitudinal temperature gradients. These rapid changes would be constrained by the lags in response of the biota (Woodward 1987, Neilson et al. 1989, Winjum and Neilson 1989). Thus, some biomes could undergo rapid changes in their boundaries as well as regional, threshold changes in physiognomy. Within biome changes are likely a combination of spatial shifts of edges of homogeneous climatic regions and within biome (climatic region) changes in seasonal amplitudes. If climatic change were to force some of these regional biomes toward thresholds of change, how fast would the change occur? What would be the nature of the changes, and would there be early warning signs?

Climatic Change – Ecotones and Thresholds

The rate of global climatic change anticipated within the next century could be extreme enough to essentially uncouple some ecosystem components from the climate system. For example, the southeast forests could be subjected to severe summer drought, and therefore cease reproduction and undergo widespread decline. The climate, primarily the thermal regime, of the forests throughout the eastern United States could shift on the order of 500–1000 km to the north within a hundred years due to global warming (Winjum and Neilson 1989). However, most heavy seeded trees can only migrate about 10–40 km per century (Zabinski and Davis 1989). Furthermore, it usually requires several decades for a tree to reach reproductive age. If we were to plant tree seedlings to the north in regions that we expect to be favorable as adults, they would likely not survive today's climate. If we plant seedlings in northern reaches that will survive today's climate, they might not be well adapted to the climate as mature individuals. In this case the disturbance is chronic, yet not from any particular stress of climate, rather from the rate of change of climate. Organisms with short life-cycles would be favored. As they reproduce and die, the climate continues to change and different species would colonize. The community (composed of short-lived organisms in any one location) might be constantly changing, but might be considered to be in equilibrium with the present, changing climate. In effect, the communities could be relegated to a semi-permanent early successional status. With a drying of the summers, the forests of the southeast, could shift to a Mediterranean (i.e., dry summer) hydrology, with possibly a savanna to woodland or some other arid component. All of the forested regions of the United States are potentially susceptible to similar, widespread physiognomic change.

Over geologic time, large-scale biogeographic changes typically occurred in a quasi-equilibrium sense for all organisms, not just those with short life-cycles. That may not be the case in the future. The rate-of-change disturbance represents a partial uncoupling of the spatial and temporal response characteristics of the climate from the response characteristics of many extant species. That is, normally slow changes over large areas could become fast changes over large areas.

Most of the regional ecological impacts from global climatic change could result from changes in the regional water balance, largely induced by increased potential evapotranspiration. These conclusions carry the assumption that the direct effects of elevated CO_2 concentration will have little to no affect on plant water use efficiency (WUE) at landscape scales. This is somewhat distinct from potential increases in WUE at the level of the leaf, when stomata are open. However, if increased WUE were to actually have an impact on regional water balance, biomes might become more drought tolerant. The widespread negative effects might not occur (Neilson et al. 1989). The direct effects of elevated CO_2 represent significant uncertainties in projecting the potential impacts of climatic change on biomes.

Monitoring to Detect Change – Ecotones and Thresholds

How would we monitor for impending changes such as those described? What would we measure to determine a spatial shift in an ecotone or to determine the proximity to a physiognomic threshold in time. Annual climatic statistics could in many cases be quite useful in signaling proximity to an ecological threshold (Fig. 1). However, changes in the seasonality will provide greater accuracy in locating climatic gradients associated with ecotones. Tracking movements of climatic gradients through space will be important. But, monitoring the response of the vegetation will be more challenging.

The size distribution of habitats as 'perceived' by plants appears to shift to smaller, more diverse patches as one moves toward an ecotone. Habitat size and variability can be measured along regional transects to determine the spatial location of the actual transition zone. The same phenomenon should occur through time, that is, the 'perceived' size distributions of habitats should shift to smaller, more diverse patches as a region shifts toward an ecological threshold (Neilson and Wullstein 1983).

The potential impacts of climatic change on habitat size and variability can be easily demonstrated with a hypothetical experiment. A hilly region, normally dry, is slowly flooded by water. The initial, homogeneous patch (from the plant 'perspective') is one of all dry land. As the water level rises, the valley bottoms would produce a distribution of small wet patches in a dry matrix. As the water rises further, the small patches merge into fewer, large and homogeneous patches. At some point these will merge into a matrix and the dry land will become isolated into ever smaller sized patches. That is, the wet-island, dry-matrix pattern reverses to a wet-matrix, dry-island pattern. Computer experiments indicate that such patterns of habitat fragmentation produce thresholds in regional ecological processes, e.g., seed dispersal or the spread and magnitude of disturbance (Gardner et al. 1987, Turner et al. 1988, Turner et al. 1989). That is, a critical level of habitat fragmentation will produce or release barriers to many horizontal processes of movement. Thresholds that alter the horizontal spread or movement of energy and material are distinct from the physiognomic threshold discussed earlier. Thus, there exist two different kinds of thresholds that could be approached as a region is subjected to climatic change: (1) physiognomic change, and (2) truncation or release of horizontal landscape processes.

History can also impart pattern to patch structure at spatial boundaries. Another example will be illustrative. Begin with the same hypothetical experiment, only impart an incline to the water table before it rises throughout the region. As the water table rises, valley bottoms on one end of the region would start to fill before those at the other end, producing a well-defined 'migrational' front of wet patches. Even though there may be valleys at the other side of the region that are lower in elevation, they will not be initially occupied by water. Similarly, the migrating front of a species' distribution tends to be sharply defined, while the trailing edge is fuzzy as it leaves behind isolated relics. As we approach temporal thresholds, there should also be predictable patterns of response by the invading versus the declining species. Such changes in patch size distribution over a whole region could be measured both remotely and from the ground. A bioassay of habitat size and variability is one approach to this measurement. Physical measurements of soil moisture through time can also indicate directional changes in patterns of habitat size and variability as regional conditions become more stressful (Neilson and Wullstein 1983). These characteristics could potentially be measured remotely and calibrated to ground data.

CONCLUSIONS

In summary, regional ecological change, induced by climatic change, could be manifested in at least two ways: 1) changes in the boundaries of regions; and 2) rapid changes in the qualities within regions. The hypothesis is presented that the concept of ecotones is equally applicable to both types of change. Ecotones in space are considered in the classical context as spatial discontinuities separating distinct ecological communities at any spatial scale, from biomes to interpatch differences. Ecotones in time are defined as ecological or physiognomic thresholds that exist throughout a region or biome and would become manifest as rapid transitions from one biotic community to another under relatively minor changes in climate.

Transitions in community composition could be significant physiognomic transitions and might be facilitated by a catastrophic shift in the regional disturbance regime. That is, if a forest, for example, were forced by drought stress to change from a closed forest condition to an open savanna, the change would likely be mediated by regional increases in disease and wildfires. As the landscape proceeds through a rapid change in habitat size and variability, thresholds in the horizontal movement of seeds, pollen and wildlife, and the spread of disturbance would likely be encountered. The potential magnitude of the spatial changes in ecotones is on the order of several hundred kilometers of horizontal movement and up to a thousand meters of vertical rise in a few decades (Neilson et al. 1989, Winjum and Neilson 1989). Such rapid changes in habitat patterns could produce the extinction of many species.

Although climatic change could, by itself, alter ecotone locations and regional ecosystem physiognomy, secondary influences such as disturbance regimes would likely modulate the location of the boundary or the timing and magnitude of the regional physiognomic change. For example, fires in a prairie could burn into an adjacent forested region shifting the ecotone from its climatically preferred position. Likewise, if a landscape is favored by climatic change to shift from a closed forest to an open savanna, a large, regional wildfire could produce a physiognomic overshoot, (e.g., a pure grassland) with a lower leaf area than the climatically 'preferred' Savanna (grass-tree mixture). That is, as the forest drys out, there should be a gradual reduction in the leaf area, toward the savanna condition. However, the fuel load could result in a large fire that might clear the landscape. The re-establishment of trees, even at a lower density could take some time, or with continuing climatic change, might not occur. Thus, the disturbance could induce a landscape change from a closed forest directly to an open grassland or shrubland, bypassing the climatically preferred savanna.

A regional monitoring program should be structured as an early warning system of such regional threshold changes. The system should use the techniques of remote sensing to provide a regional perspective, but should be calibrated to measurable, local-scale patterns. The concept of regional control of local habitat patch size is presented as providing a potential signature of proximity to an ecotone, either in space or time and the temporal dynamics of that ecotone. Patches are hypothesized to be large in core regions and small and diverse near transition zones. To determine the scale of habitat patches and detect changes in that scale along regional and temporal gradients, monitoring programs will require careful design. Managers should address these design considerations in current planning efforts. They should also consider the prospect of regional ecological thresholds and the impacts of different management practices when confronting rapid change, increased disturbance regimes, new species mixes and altered landscape patterns. Careful management of disturbance regimes through manipulations of fuel load or cutting practices, for example, will be important in

minimizing both the magnitude of ecological changes and the rate of the changes over regional extents. A combination of purely observational data and bioassay approaches (i.e., placement of organisms in the environment as a bioassay of habitability) should be considered when monitoring for ecological change. Monitoring indicators should be developed to provide useful information at a range of scales from individual organism responses to landscape patterns.

ACKNOWLEDGEMENTS

The ideas presented in this paper represent a synthesis of separate research projects spanning several years. I would like to thank the many scientists and students who influenced my thinking and provided encouragement. Recent work was accomplished with the valuable assistance of George King, Robert DeVelice and Jim Lenihan. The author thanks Robert Gardner, Marjorie M. Holland, Mari N. Jensen, Robert J. Naiman and an anonymous reviewer for reviews of earlier drafts of this manuscript. This research was funded, in part, by a Cooperative Agreement from the Environmental Protection Agency to Ronald P. Neilson (CR814620).

LITERATURE CITED

Allen, T. F. H., and T. B. Starr. 1982. Hierarchy: perspectives for ecological complexity. University of Chicago Press, Chicago, Illinois, USA.

Axelrod, D. I. 1967. Drought, diastrophism, and quantum evolution. Evolution **21**:201–209.

Brown, J.H., and A.C. Gibson. 1983. Biogeography. C.V. Mosby Company, St. Louis, Missouri, USA.

Burke, M. J., L. V. Gusta, H. A. Quamme, C. J. Weiser, and P. H. Li. 1976. Freezing injury in plants. Annual Review of Plant Physiology **27**:507–528.

Cody, M. L. 1975. Towards a theory of continental species diversity. Pages 214–250 in M. L. Cody and J. M. Diamond, editors. Ecology and evolution of communities. Belknap Press, Cambridge, Massachusetts, USA.

Currie, D. J., and V. Paquin. 1987. Large-scale biogeographical patterns of species richness of trees. Nature **329**:326–327.

Delcourt, P. A., and H. R. Delcourt. In press. Ecotone dynamics in space and time. In F. di Castri and A. J. Hansen, editors. Landscape boundaries: consequences for biotic diversity and ecological flows. SCOPE book series. Springer–Verlag, New York, New York, USA.

di Castri, F., A. J. Hansen, and M. M. Holland, editors. 1988. A new look at ecotones: emerging international projects on landscape boundaries. Biology International, Special Issue **17**:1–163.

Dice, L. R. 1943. The biotic provinces of North America. University of Michigan Press, Ann Arbor, Michigan, USA.

Gardner, R. H., B. T. Milne, M. G. Turner, and R. V. O'Neill. 1987. Neutral models for the analysis of broad–scale landscape pattern. Landscape Ecology **1**:19–28.

Küchler, A. W. 1964. The potential natural vegetation of the conterminous United States. American Geographical Society Special Publication Number 36, New York, New York, USA.

MacArthur, R.H. 1972. Geographical ecology: patterns in the distribution of species. Harper & Row, New York, New York, USA.

Mooney, H. A. 1977. Convergent evolution in Chile and California. Mediterranean climate ecosystems. US/IBP Synthesis Series 5. Dowden, Hutchinson & Ross Publishing Company, Stroudsburg, Pennsylvania, USA.

Neilson, R. P. 1986. High–resolution climatic analysis and southwest biogeography. Science **232**:27–34.

Neilson, R. P. 1987a. Biotic regionalization and climatic controls in western North America. Vegetatio **70**:135–147.

Neilson, R. P. 1987b. On the interface between current ecological studies and the paleobotany of pinyon–juniper woodlands. Pages 93–98 in R. L. Everett, editor. Proceedings of the Pinyon–Juniper Conference. United States Department of Agriculture Forest Service General Technical Report INT–215, Ogden, Utah, USA.

Neilson, R. P., G. A. King, R. L. DeVelice, J. Lenihan, D. Marks, J. Dolph, W. Campbell, G. Glick. 1989. Sensitivity of ecological landscapes and regions to global climatic change. United States Environmental Protection Agency, EPA–600–3–89–073, NTIS–PB90–120–072–AS, Washington, D.C., USA.

Neilson, R. P., and L. H. Wullstein. 1983. Biogeography of two southwest American oaks in relation to atmospheric dynamics. Journal of Biogeography **10**:275–297.

Neilson, R. P., and L. H. Wullstein. 1986. Microhabitat affinities of Gambel Oak seedlings. Great Basin Naturalist **46**:294–298.

Omernik, J. M. 1987. Ecoregions of the conterminous United States. Annals of the Association of American Geographers **77**:118–125.

O'Neill, R. V., D. L. DeAngelis, J. B. Waide, and T. F. H. Allen. 1986. A hierarchical concept of ecosystems. Princeton University Press, Princeton, New Jersey, USA.

Peet, R. K. 1978. Latitudinal variation in southern Rocky Mountain forests. Journal of Biogeography **5**:275–289.

Pielou, E. C. 1975. Ecological diversity. Wiley & Sons, New York, New York, USA.

Quinlan, F. T. , T. R. Karl, and C. N. Williams, Jr. 1987. United States historical climatology network serial temperature and precipitation data. Carbon Dioxide Information Analysis Center NDP–019, Oak Ridge National Laboratory, Oak Ridge, Tennessee, USA.

Ricklefs, R.E. 1987. Community diversity: relative roles of local and regional processes. Science **235**:167–171.

Simpson, G.G. 1964. Species density of North American recent mammals. Systematic Zoology **13**:57–73.

Stebbins, G. L., Jr. 1952. Aridity as a stimulus to plant evolution. The American Naturalist **86**:33–44.

Stephenson, N. L. 1990. Climatic control of vegetation distribution: the role of the water balance. American Naturalist **135**:649–670.

Tibbits, T.H. 1979. Humidity and plants. Bioscience **29**:358–363.

Turner, M.G., R.H. Gardner, V.H. Dale, and R.V. O'Neill. 1988. Landscape pattern and the spread of disturbance. Pages 373–382 *in* M.R.T. Hrniciarova and L. Miklos, editors. Proceedings of 8th International Symposium on problems of landscape ecological research. Institute of Experimental Biology and Ecology, Bratislava, Czechoslovakia.

Turner, M.G., R.H. Gardner, V.H. Dale, and R.V. O'Neill. 1989. Predicting the spread of disturbance across heterogeneous landscapes. Oikos **55**:121–129.

United States Department of Agriculture. 1941. Climate and Man: Yearbook of Agriculture. House Document Number 27, 77th Congress, 1st Session, Washington, D.C., USA.

US WEST. 1988. Hydrodata User's Manual: United States Geological Survey Daily and Peak Values, Version 2 Edition. US WEST Optical Publishing, Denver, Colorado, USA.

Watson, M. A. 1980. Shifts in patterns of microhabitat occupation by six closely related species of mosses along a complex altitudinal gradient. Oecologia **47**:46–55.

Whitford, W. G., J. F. Reynolds, and G. L. Cunningham. 1987. How desertification affects nitrogen limitation of primary production on Chihuahuan Desert watersheds. Pages 143–153 *in* E. F. Aldon, C. E. Gonzales-Vincente, and W. H. Moir, editors. Strategies for classification and management of native vegetation for food production in arid zones. United States Department of Agriculture Forest Service General Technical Report RM–150, Fort Collins, Colorado, USA.

Whittaker, R. H. 1972. Evolution and measurement of species diversity. Taxon **21**:213–251.

Whittaker, R.H. 1975. Communities and ecosystems. Macmillan Publishing Company, New York, New York, USA.

Whittaker, R. H., and S. A. Levin. 1977. The role of mosaic phenomena in natural communities. Theoretical Population Biology **12**:117–139.

Winjum, J.K., and R.P. Neilson. 1989. The potential impact of rapid climatic change on forests in the United States. Pages 71–92 *in* J. B. Smith and D. A. Tirpak, editors. The potential effects of global climate change on the United States. United States Environmental Protection Agency, EPA–230–05–89–050, Washington, D.C., USA.

Woodward, F. I. 1987. Climate and plant distribution. Cambridge University Press, London, England.

Zabinski, C., and M. B. Davis. 1989. Hard times ahead for Great Lakes forests: a climate threshold model predicts responses to CO_2-induced climate change. Pages 1–19, Appendix D, Chapter 5, in J. B. Smith and D. A. Tirpak, editors. The potential effects of global climate change on the United States. United States Environmental Protection Agency, EPA–230–05–89–054, Washington, D.C., USA.

POTENTIAL RESPONSES OF LANDSCAPE BOUNDARIES TO GLOBAL ENVIRONMENTAL CHANGE

MONICA G. TURNER, ROBERT H. GARDNER, AND ROBERT V. O'NEILL. Environmental Sciences Division, Oak Ridge National Laboratory, P.O. Box 2008, Oak Ridge, Tennessee 37831-6038, USA.

Abstract. Global change is likely to affect the location, size, shape, or composition of landscape boundaries. Neutral models were used to study two general mechanisms by which landscape boundaries may respond to global change: (1) disturbance regimes may change in response to climate, leading to rapid alterations in landscape structure, and (2) in the absence of disturbance, suitable habitat for different species may move gradually and directionally. The spread of disturbance was simulated as a function of the proportion of the landscape occupied by a disturbance-prone habitat and the frequency (probability of initiation) and intensity (probability of spread) of a habitat-specific disturbance. The effects of changing disturbance regimes on landscape boundaries were different in connected and fragmented landscapes. In connected landscapes, an increase in disturbance intensity caused landscape boundaries to decrease. In landscapes that were fragmented, an increase in disturbance frequency resulted in a decrease in landscape boundaries. The qualitative change in sensitivity to disturbance intensity and frequency occurred when the proportion of disturbance-susceptible habitat was near the critical threshold, p_c, which is approximately 0.6 in random landscapes. Habitat displacement and species migration were simulated as a function of the proportion of available landscape occupied by a community, migration (probability of colonizing adjacent cells), extinction (probability of the community becoming locally extinct), and the rate at which potential habitat (i.e., climatic conditions suitable for establishment) is displaced. The simulated community could track the movement of habitat when the displacement rate was slow, the extinction probability was low, and the migration probability was at least moderate. Spatial lags (distance of remnants from main front) were greatest when extinction was low and the habitat moved quickly. Simulations such as these can be used to identify combinations of parameters which allow successful colonization and those for which community survival becomes critical. Neutral models can generate quantitative, testable hypotheses which can be used to better understand and predict landscape responses to global change.

Key words: climate change, landscape ecology, disturbance, species migration, boundaries.

INTRODUCTION

Dramatic changes in the global environment are predicted to occur within the next century. For example, the projected doubling of atmospheric CO_2 is expected to alter global climate. Current predictions are for an average rise in global temperature of 1.5 to 4.5°C, with greater warming in winter than summer and increased warming with increased latitude. Increased precipitation in high latitudes and decreased summer precipitation and soil moisture in middle latitudes of the northern hemisphere are also predicted. However, the magnitude, rate, and spatio-temporal characteristics of the climatic response are uncertain. Given the strong relationship between climate, atmosphere, soils, biota, and human activities, there is a solid basis for anticipating changes in terrestrial biomes in response to changes in the global environment.

Global change is likely to have substantial effects on landscape structure and function. Ecological processes (e.g., plant succession, nutrient cycling, dispersal, and disturbance) occur in a spatial context and interact with spatial abiotic factors (e.g., topography and soils) to produce complex patterns on the landscape. These spatial patterns constrain the biota. For example, suitable habitat and resources for different organisms may be delimited by landscape patterns. In turn, biotic interactions may generate or alter the spatial patterns. This interplay between pattern and process is the key focus of landscape ecology (e.g., Risser et al. 1984; Forman and Godron 1986; Turner 1987, 1989; Urban et al. 1987).

The spatial patterning of landscapes implies the existence of boundaries or ecotones between adjacent ecological systems. Boundaries may influence landscape dynamics by exerting control over the flow of energy, materials, and organisms between landscape components (Risser et al. 1984, Wiens et al. 1985, Forman and Godron 1986, Hansen et al. 1988). It has been suggested that research on ecotones should be prominent in exploring the consequences of global change because interactions between various landscape components will often occur at these boundaries (Hansen et al. 1988).

Landscape boundaries may change in location, size, shape, or composition in response to global change. Specific responses will depend upon a variety of factors at many spatial and temporal scales (e.g., edaphic conditions, local weather patterns, present vegetation mosaic and species composition). However, two general mechanisms by which landscape boundaries may change with the global environment can be considered: (1) disturbance regimes may change in response to climate, leading to rapid alterations in landscape structure; or (2) in the absence of disturbance, suitable habitat for different species may move gradually and directionally. A gradual change in potential habitat may lead to a delayed change in actual landscape boundaries if species' responses exhibit significant time lags. In this paper, we discuss these two mechanisms and explore the potential responses of landscape boundaries to global change by using a neutral modeling approach (sensu Caswell 1976, Gardner et al. 1987). "Neutral models" are used to generate an expected result in the absence of a site-specific process or pattern.

Changing Disturbance Regimes

The patterns of many landscapes are influenced or controlled by disturbance (e.g., White 1979, Mooney and Godron 1983, Pickett and White 1985, Turner 1987, Baker 1989). For example, disturbances create openings within forested landscapes, leading to mosaics of successional patches of different ages (e.g., Runkle 1985, Knight 1987, Baker 1989). Landscape patterns in old-growth forests of New England result from frequent natural disturbances, such as windstorms, lightning, pathogens, and fire (Foster 1988a). The diversity of the Yellowstone National Park landscape is controlled by a natural fire regime in which catastrophic fires recur at 200- to 300-year intervals (Romme 1982, Romme and Knight 1982). The susceptibility of individual sites to disturbance may be controlled by edaphic conditions such as slope position and aspect (Foster 1988a), and therefore disturbances vary spatially. The spatial patterns of disturbances such as wind damage (Foster 1988b) or fire (Sellers and Despain 1976, Romme 1982, Romme and Knight 1982) also may be predictable based on the arrangement of age classes of the vegetation.

The frequency, duration, and severity of both abiotic and biotic disturbances are likely to be altered by climate change. For example, forest fire frequencies should increase where the climate becomes warmer and drier (Sandenburgh et al. 1987). Forest fire regimes may also change in response to species composition shifts induced by elevated CO_2 and climate change. Patterns of biotic disturbances may also be altered. Because their ranges are often limited by climatic factors, the distributions of pests or pathogens may change with climate. For example, spruce budworm populations may expand their range under a warmer climate because they are favored by warm, dry, early-spring conditions (Haynes 1982).

If ecological disturbance regimes are altered, changes are likely in many landscapes and, therefore, in landscape boundaries. Disturbance regimes can be described by a variety of characteristics, including spatial distribution, frequency, return interval, rotation period, predictability, area, intensity, severity, and synergism (e.g., Rykiel 1985, White and Pickett 1985). In this study, we use simulation models to examine changes in two disturbance characteristics, intensity and frequency, and their interactions with landscape pattern. We define disturbance frequency as the probability that a new disturbance will be initiated in a unit of susceptible habitat during the time period represented by the simulation. Intensity is defined as the probability that the disturbance, once initiated, will spread to adjacent cells of the same habitat. We predict the spread of a disturbance across a landscape as a function of (1) the proportion of the landscape occupied by a disturbance-prone cover type, (2) disturbance intensity, and (3) disturbance frequency.

SIMULATION EXPERIMENTS

Two-dimensional m x m landscape maps (100 x 100 cells) were randomly generated for the disturbance-susceptible habitat at different values of p. The variable, m, represents the number of rows or columns in the square matrix. The probability, p, for maps this large represents the proportion of a landscape occupied by a susceptible habitat. The remainder of the landscape is considered unsuitable for propagation of the simulated disturbance. Thus, the maps are composed of 10,000 cells, each of which is randomly designated as suitable or unsuitable for a disturbance. A disturbance that could propagate through susceptible habitat was characterized by frequency and intensity. Disturbance frequency, f, is the probability that a new disturbance will be initiated in a unit of susceptible habitat during the simulation (e.g., the probability that lightning striking a hectare of pine forest during a particular storm event or time period will ignite a fire). Disturbance intensity, i, is defined as the probability that the disturbance, once initiated, will spread to adjacent cells of the same habitat (e.g., the probability of fire or a pathogen spreading to an adjacent cell of susceptible forest). In the first set of simulations (n = 10 replicates), frequency (f = 0.01, 0.1, 0.5), intensity (i = 0.25, 0.5, 0.75), and habitat probability (p = 0.4, 0.8) were all varied. In random landscapes, low values of p yield landscapes with small isolated patches, whereas high values of p yield large connected patches (Gardner et al. 1987). This set of simulations was designed to give a general impression of the interplay of p, i, and f. In a second set of simulations (n = 10), we assumed that disturbances (e.g., lightning strikes or pest outbreaks) would occur at some unavoidable minimum number. Therefore, frequency was held constant at f = 0.01. This permitted a more complete examination of the interplay of the controllable parameters, p and i. In a forest, for example, p could be controlled by old-growth cutting policies and i could be controlled by containing fires.

The boundaries in the landscape were analyzed for each map before disturbances began. Boundary length was measured by using the side of one grid cell as a unit length. Landscape disturbance was then simulated as follows. Cells were randomly disturbed at a given frequency until exactly fpm^2 disturbances were initiated (m^2 is equal to the size of the map, in this case 10,000). Each disturbed cell was then changed to a state that was no longer vulnerable to disturbance. The disturbance was then randomly propagated with an intensity, i, to an adjacent disturbance-prone cell. The propagation process was repeated until the disturbance could not spread any farther. The disturbed landscape was then analyzed, and the pattern and boundaries between disturbed and undisturbed habitat were summarized.

For this analysis, we distinguish among different types of landscape boundaries that may have ecological significance. We define "habitat boundary" as the edges separating susceptible and nonsusceptible habitat (e.g., the boundaries of forest patches susceptible to a pest). We define "disturbance boundary" as edges between disturbed cells and either undisturbed or nonsusceptible cells (e.g., the boundaries of the infested patches). For both habitat and disturbance boundaries, we distinguish between inner and outer edges. An outer edge refers to the outermost perimeter of a cluster of undisturbed (outer habitat edge) or disturbed (outer disturbance edge) habitat. An inner edge is next to a "clearing" or gap within a patch (e.g., inner habitat edges would surround a gap in a patch of undisturbed habitat). Inner disturbance edges surround undisturbed islands that are embedded in a larger disturbed area.

RESULTS

Interaction of disturbance frequency (f) and intensity (i) in fragmented and connected landscapes

Increasing disturbance intensity (*i*) reduces the total amount of habitat boundaries in the landscape (Table 1). Greater intensity leads to larger continuous patches of disturbance and, therefore, fewer habitat boundaries. In contrast, increasing disturbance intensity almost always increases total length of disturbance boundaries. The greater the spread of the disturbance, the greater the opportunity for disturbance boundaries to form. The only exception is for high *p* (e.g., *p* = 0.8) and high *f* (e.g., *f* = 0.5). Under these conditions, the high frequency of disturbance creates many small disturbance patches, and further spread of the disturbance causes the small disturbances to coalesce into larger contiguous areas with less boundary length. When *p* and *f* are high, the percentage of boundary that is inner edge (in parentheses in Table 1) indicates that most of the total boundary length surrounds isolated habitat patches embedded in the disturbed area.

Inner edges are not a significant feature of fragmented (e.g., *p* = 0.4)

landscapes (Table 1). When the susceptible habitat is scattered on the landscape, there are very few isolated gaps within patches of susceptible habitat, and this pattern is unaffected by disturbance (Turner et al. 1989). However, inner habitat edges become more important when the susceptible habitat covers most of the landscape (e.g., *p* = 0.8). Inner habitat edges increase with disturbance if the disturbance is of low frequency (e.g., *f* = 0.01, 0.1) and low to medium intensity (e.g., *i* = 0.25, 0.5). Under these circumstances, many small gaps are formed, and the inner edges may form the majority of the total landscape boundaries. For disturbances of relatively low frequency (*f* = 0.01), the ratio of inner to outer habitat edges is maximized for high values of *p* (continuous cover) combined with low values of intensity, *i* (small pockets of disturbance) (Fig. 1a). However, inner disturbance edges increase with disturbance intensity at all levels of frequency in very connected landscapes (e.g., *p* = 0.8). The more the disturbance spreads, the greater the likelihood that small habitat patches will become isolated within the large disturbed areas. The ratio of inner to outer disturbance edges increases as *p* and *i* increase (Fig. 1b).

Boundaries as a function of the probability of susceptible habitat (p) and disturbance intensity (i)

We now consider the interaction of *p* and *i* when a fixed low frequency of disturbance (*f* = 0.01) is assumed. The total habitat boundaries as a function of *p* and *i* are shown in Figure 2. Maximum habitat boundaries occur at low to intermediate values of *p* (dispersed landscapes) and lower values of *i* (small disturbances). Under these conditions, the landscape is composed of many small patches with many boundaries. Habitat boundaries on dispersed landscapes are relatively insensitive to changes in disturbance intensity. Because the patches are small and isolated, higher intensity disturbances still tend to be contained within the small patches and have little effect on total boundary. Above a critical threshold

TABLE 1. Mean number (**n** = 10) of total habitat and disturbance boundaries for various values of the proportion of susceptible habitat, **p**; disturbance intensity, **i**; and disturbance frequency, **f**. Values in parentheses are the percentage of the total edge composed of inner (isolated) edges.

Disturbance intensity, *i*	Habitat Boundaries		Disturbance Boundaries	
	p = 0.4	**p** = 0.8	**p** = 0.4	**p** = 0.8
		Disturbance frequency **f** = 0.01		
0.25	9644 *	6885 (90)	205 *	558 *
0.50	9554 *	6708 (78)	286 *	1947 (10)
0.75	9346 *	1999 *	444 *	6269 (88)
		f = 0.1		
0.25	8948 *	8256 (69)	1858 *	4284 *
0.50	8250 *	5739 (6)	2400 *	6169 (43)
0.75	6986 *	1289 *	3281 *	6426 (89)
		f = 0.5		
0.25	5255 *	6243 *	6776 *	8622 (57)
0.50	3870 *	2504 *	7193 *	7222 (87)
0.75	2559 *	511 *	7698 *	6623 (90)

* An asterisk indicates that inner edges are less than 2% of the total.

RATIO OF INNER/OUTER EDGES

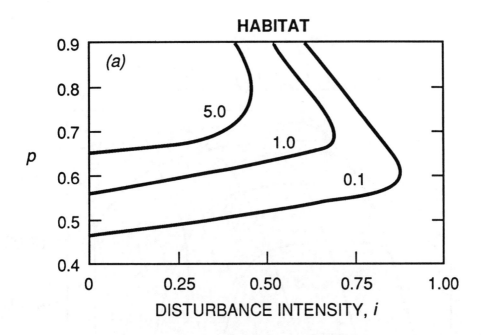

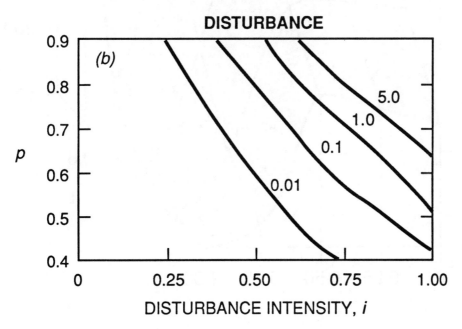

Figure 1. Ratio of inner/outer edges for (a) habitat boundaries and (b) disturbance boundaries as a function of disturbance intensity (i) and the probability of susceptible habitat (p). The figures show isopleths along which the ratio shows the same value.

TOTAL HABITAT EDGES

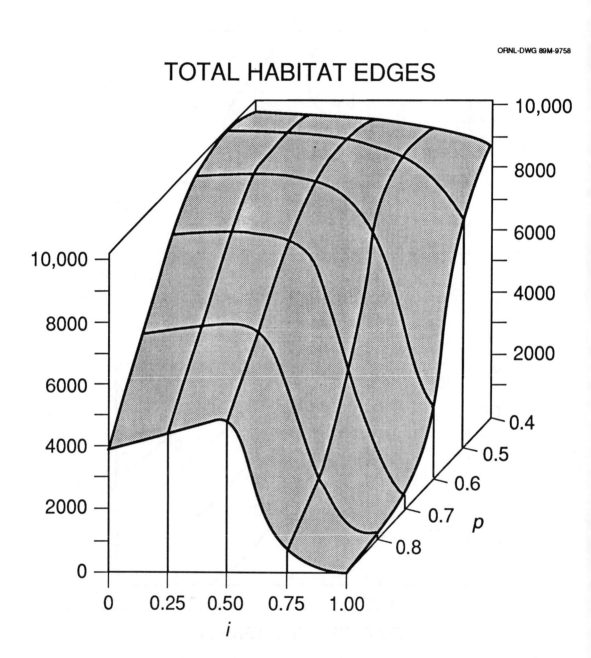

Figure 2. Total habitat boundaries as a function of disturbance intensity (i) and the probability of susceptible habitat (p) for fixed, low-frequency disturbances (f = 0.01).

($p_c \sim 0.6$), patches in random landscapes tend to be connected (Gardner et al. 1987), and habitat boundaries are increased by disturbances of moderate intensity (Fig. 2). (However, this critical threshold of connectivity is a function of the shape of the grid [see Stauffer 1985]). Moderate intensity disturbance breaks up the tight patch structure and creates more edges. The greatest sensitivity to disturbance occurs as p and i increase together (Fig. 2). When this occurs, the landscape is covered by a few large clusters which coalesce at high intensity disturbances.

The total disturbance boundaries as a function of p and i are shown in Figure 3 (note that the p and i axes are rotated from their position in Fig. 2). Disturbance boundaries are insensitive to i on dispersed landscapes (low p) and insensitive to p at low disturbance intensities. In both of these cases, there is very little disturbance edge and therefore very little change with a change in p or i. Maximum disturbance boundaries (see Table 3) tend to occur at higher intensities and at intermediate levels of cover ($0.6 < p < 0.9$).

Maximum outer habitat edges only occur on dispersed landscapes (Fig. 4a). Above p_c, intermediate intensities break up the large clusters and create more outer edges than were present when the landscape was not disturbed. In contrast, maximum inner habitat edges occur only at the lower levels of intensity and intermediate values of p ($0.6 < p < 0.9$) (Fig. 4c). Under these conditions, the disturbances create many small clearings within contiguous habitat clusters.

Maximum values of outer disturbance edges (Fig. 4b) are found along a diagonal running from $p = 0.9$, $i = 0.4$ to $p = 0.6$, $i = 1.0$. As one deviates from this line, conditions become less favorable for producing outer disturbance edges. A minimum occurs when both p and i are very high (disturbances completely cover the large patches, leaving few boundaries of any kind). A second minimum occurs on more dispersed

landscapes at low intensity disturbances. Under these circumstances, habitat boundaries are favored rather than disturbance boundaries (see Figs. 2 and 3). Inner disturbance edges are maximized (Fig. 4d) at high intensity and intermediate cover ($0.6 < p < 0.9$). This combination maximizes the probability of an extensive disturbed area surrounding and enclosing undisturbed habitat.

Maximum Landscape Boundaries

In the absence of disturbance, maximum habitat edges on a random landscape occur when $p = 0.5$. An edge is simply the juxtaposition of a susceptible cell (probability p) and a nonsusceptible cell [probability $(1 - p)$]. The probability, E, of a cell being located on a habitat edge is simply the product of these two probabilities:

$$E = p(1 - p). \quad (1)$$

Recalling from elementary calculus that a function has its maximum or minimum at the zero of its first derivative, we can take the first derivative of (1) with respect to p,

$$d/dp\,[p(1 - p)] = 1 - 2p \quad (2)$$

Setting the right hand side of Eq (2) equal to zero and solving for p, we find $p = 0.5$. Again recalling that the solution to Eq (2) is a maximum if the second derivative of Eq (1) is negative, we find that $p = 0.5$ maximizes the probability of edges.

The situation changes slightly as we add a point disturbance (i.e., disturbance intensity $i = 0.0$ and the disturbance does not spread). Now an edge is the juxtaposition of remaining susceptible habitat [probability $p(1-f)$] with either nonsusceptible habitat [probability $(1 - p)$] or disturbed habitat (probability fp). The probability of a random point being an edge becomes:

$$p(1-f)(1-p) + p(1-f)(fp) = p(1-f)(1-p+fp) \quad (3)$$

The zero of the first derivative with respect to p of Eq(3) for $f \neq 1$ (because we are not

59

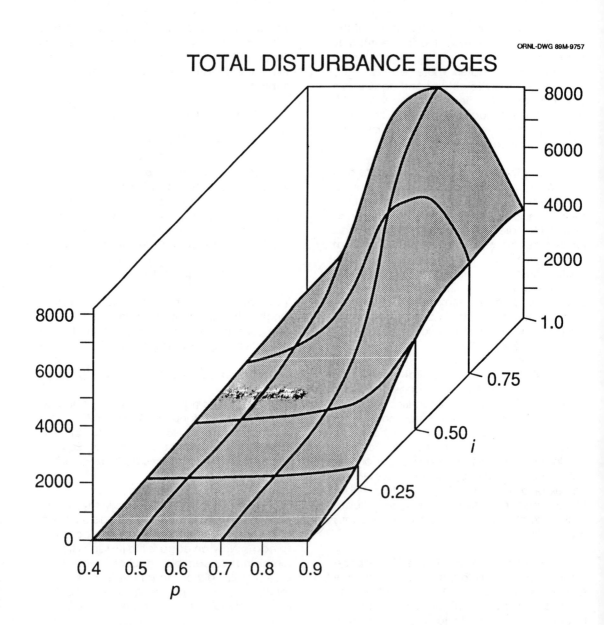

TOTAL DISTURBANCE EDGES

Figure 3. Total disturbance boundaries as a function of disturbance intensity (i) and the probability of susceptible habitat (p) for fixed, low-frequency disturbances (f = 0.01).

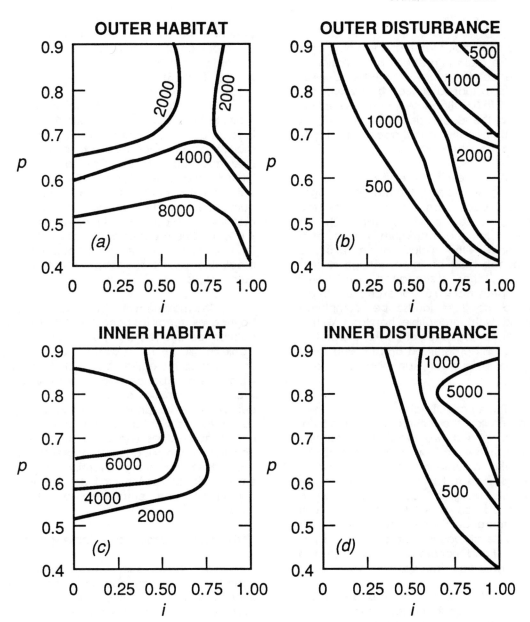

Figure 4. Isopleths showing equal amounts of edge as a function of the probability of susceptible habitat (p) and disturbance intensity (i) for (a) outer habitat edges, (b) outer disturbance edges, (c) inner habitat edges, and (d) inner disturbance edges. Disturbance frequency was fixed at f = 0.01.

interested in the singularity when all cells are disturbed) becomes:

$$1 - 2p + 2fp = 0 \quad (4)$$

Equation (4) indicates that the value of **p** giving the maximum edges is a function of disturbance frequency, **f**. Table 2 gives values for **p** that satisfy Eq (4) at values of **f** from 0.01 to 0.5. Because **f** < 1.0 for practical situations, the second derivative of Eq (3) will always be negative and the solution of Eq (5) will be a maximum.

As the frequency of point disturbances increases, the maximum edges are achieved on more connected landscapes, i.e., **p** > 0.5 (Table 2). In essence, for the maximum edges after disturbance, the value of **p** prior to disturbance must be large enough so that following the disturbance, the final landscape has **p** = 0.5. For example with **p** = 1.0, we have a landscape completely covered with susceptible habitat. With a disturbance frequency of **f** = 0.5, one half of the cells will be disturbed and the final result is the same as having an initial landscape with **p** = 0.5.

The situation becomes more complex when we allow the disturbance to spread at intensity, **i**, and we want to distinguish between inner and outer edges. Table 3 shows the disturbance intensity at which maximum habitat and disturbance boundaries occur for each value of **p** when disturbance frequency is low (**f** = 0.01). Even with disturbance spread, maximum habitat boundaries still occur near **p** = 0.5 for small values of **f** and monotonically decrease as **p** moves away from that value.

On dispersed landscapes (**p** = 0.4, 0.5) the maximum habitat edges are found at **i** = 0.0 (i.e., the point disturbances). Higher disturbance intensities simply open up more contiguous areas of disturbance and decrease the total habitat edge. As **p** approaches 0.6, we know from percolation theory (Gardner et al. 1987) that the habitat clusters tend to coalesce. As the habitat

patches become more contiguous, disturbances of moderate intensity break up the patches and create more edge. The disturbance intensity that creates the most boundary increases from **i** = 0.25 to **i** = 0.50 as proportion of the landscape occupied by susceptible habitat increases from **p** = 0.6 to **p** = 0.9 (Table 3).

The values in parentheses in Table 3 indicate the percent of the boundary that is inner edge. On dispersed landscapes, the majority of the edges are always outer edges (see also Table 1 and Fig. 1a). But at intermediate values of **p** (e.g., 0.6 to 0.8), the disturbances occur in the center of contiguous patches of habitat and the majority of the edges are inner edges. At **p** = 0.9, maximum edges occur at **i** = 0.5. At this higher intensity, the disturbance tends to break through the enclosing habitat to form outer edges.

The maximum number of disturbance boundaries tend to occur at the highest disturbance intensity (**i** = 1.0) when **f** = 0.01 (Table 3). At **i** = 1.0, a disturbance spreads until it covers the entire habitat cluster within which the disturbance begins. It is only at **p** = 0.9 that intensities above **i** = 0.6 tend to disturb the entire landscape, leaving few or no edges. As with habitat boundaries, inner disturbance edges (values in parentheses) dominate when **p** is equal to or greater than 0.6. In these cases, high intensity causes the entire habitat patch to be disturbed, surrounding and entrapping undisturbed areas.

Gradual Movement of Suitable Habitat

We have seen how landscape boundaries may change in response to alterations in disturbance frequency and intensity. Landscape boundaries also may change in response to interactions between global change and habitat displacement. As climate changes, for example, there may be a gradual movement of potentially suitable conditions for different species, and species may migrate to track hospitable environments. However, migration rates are

TABLE 2. Values of the proportion of susceptible habitat (*p*) that produce maximum habitat boundaries when point disturbances (i.e., disturbance intensity = 0) occur at different levels of disturbance frequency (*f*).

Disturbance frequency *f*	Proportion of susceptible habitat *p*
0.00	0.50
0.01	0.50
0.1	0.56
0.2	0.62
0.3	0.71
0.4	0.83
0.5	1.00

TABLE 3. The value of disturbance intensity, *i*, that creates maximum habitat and disturbance boundaries at *f* = 0.01 for different values of *p*. Numbers in parentheses represent the percentage of the maximum that is composed of inner edges.

p	Maximum habitat boundaries		Maximum disturbance boundaries	
	Disturbance intensity *(i)*	Amount	Disturbance intensity *(i)*	Amount
0.4	0.0	9709 (2)	1.0	858 (5)
0.5	0.0	10110 (11)	1.0	2757 (20)
0.6	0.25	9750 (52)	1.0	7397 (71)
0.7	0.25	8718 (85)	1.0	8128 (91)
0.8	0.40	6906 (87)	1.0	6575 (91)
0.9	0.50	4861 (45)	0.6	4669 (93)

difficult to predict because the projected rate of climate change is an order of magnitude faster than previous climate changes. Davis (*in press*) suggests that the spatial displacement of habitats can be visualized if one assumes the same general configuration of climate as today. With a latitudinal lapse rate in the Great Lakes region of about 100 km/°C, isotherms would be displaced northward 300 km in the next 100 years. This rate is 10 times greater than the documented range extensions for trees in a similar time interval (Davis *in press*). Furthermore, there are now new barriers to migrations (e.g., cities, agriculture, and roads) and new modes of migration (e.g., cars, trains, transplants for horticulture, forestry, or agriculture.) Range extension in the future may be less efficient than in the past because advance disjunct colonies have been extirpated by human disturbances, and propagule sources have often been reduced (Davis 1989). The current spatial distribution and abundance of a species will influence its ability to migrate successfully to regions of suitable climate and soils (Peters and Darling 1985).

A variety of modeling approaches have been used to explore the potential redistribution of habitats in response to changes in climate and atmospheric CO_2 (e.g., Davis and Botkin 1985; Emanuel et al. 1985*a*, 1985*b*; Solomon and Webb 1985; Solomon 1986). Diffusion models have been used to explore alternative modes of species migration in which (1) species may migrate as a continuous front, or (2) discontinuous populations may be established far in advance of the main species front (Davis 1987). Individual-based forest stand models have been used to simulate responses to global climate change (e.g., Davis and Botkin 1985, Solomon 1986, Pastor and Post 1988), although species dispersal/migration is not included. Simulation results suggest that species abundances are not always in equilibrium with climate; biotic interactions or other environmental factors (e.g., soil heterogeneity) can have strong effects and may obscure or delay observed responses;

and time lags of up to 1000 years may occur in biome shifts.

The rate of movement of potential habitat and the rate of species dispersal/migration will influence landscape boundaries, especially if time lags occur (e.g., Davis 1984, Webb 1986, Pennington 1986). Time lags in species responses to historical climatic changes have been documented. For example, beech has animal-dispersed seeds and tends to move as a front. Beech showed a time lag of 500 to 1000 years in crossing from the eastern to the western shore of Lake Michigan (Davis 1989). In contrast, hemlock, whose wind-dispersed seeds can travel 100 km beyond the main species front, showed no time lags attributable to crossing the Great Lakes (Davis 1989). Davis (1989) also proposes that time lags in the adjustments of species abundances to climate will be small (~10 - 20 yrs) in the heavily disturbed communities that cover most of the landscape. The most common species will be dispersed to new habitats by humans, but time lags will be a problem for unmanaged forests, natural areas, and preserves. The climate change during the next century may not be enough to kill dominant species directly. Thus, the landscape may appear superficially unchanged even though a different community is becoming established in the understory.

The migration of species and the redistribution of community types in response to global change and the associated effects on landscape boundaries are complex issues (c.f., Brubaker 1986, Pennington 1986). Some organisms track climate change closely, reacting to conditions each year, while others respond so slowly that only long-term climatic trends have any observable impact (Davis 1984). In a new set of simulation experiments, we examine potential changes in landscape boundaries by focusing on: (1) the rate of movement of conditions that are potentially suitable for a community type; (2) the probability of local extinction of the community; and (3) the probability of migration. The term "community" is used here to represent the group of organisms of interest, but the model can be developed for a species or population. We use a model very similar to the one described previously, but we modify the parameter i and add new parameters. We define the interval of movement of potential habitat, t, as the number of time steps required for a new row of the landscape matrix to become potentially available for colonization while an old row at the bottom becomes unsuitable for new colonization. Extinction, e, is defined as the probability of the community becoming extinct at a particular cell and is based upon the longevity of the community. For example, a community that exhibits a longevity, l, of 10 years would have $e = 1/10 = 0.1$. This inverse relationship is similar to turnover time and turnover rate. Migration, i, is defined as the probability of successfully colonizing an adjacent cell of potential habitat. Migration can occur in any of the four cardinal directions (up, down, left, right) from an occupied cell. Thus, the spread of the community is simulated as a continuous front, rather than as establishment of disjunct colonies.

SIMULATION EXPERIMENTS

Two-dimensional landscape maps (100 x 100 cells) were randomly generated to contain the community of interest only in the lower half of the map (e.g., rows 51 to 100) at probability p. Initially, the upper half of the map is considered to be unsuitable for the community. Spatial patterns and boundaries are analyzed on the initial landscape. At each subsequent time step during the simulation, a community cell can become extinct at probability e, and new community cells can become established within the suitable range (e.g., initially only within rows 51 to 100) at migration probability i. The movement of potential habitat (i.e., suitable range) is then simulated as follows. At each interval of time t, the lower-most row of potential habitat becomes unsuitable for the establishment of new community cells, and a new row of potential habitat becomes available at the upper edge of the range (Fig. 5). Community extinction and

establishment continue, and the range of suitable habitat continues to move at intervals of t. The simulation is concluded when the suitable habitat has been displaced to the other half of the map (i.e., rows 1 to 50). The new landscape patterns are then analyzed.

Communities may exhibit spatial and temporal lags if habitat displacement is rapid relative to the longevity of the community. This occurs because long-lived remnant populations may persist even though the habitat is no longer suitable for reestablishment. As a result, landscape patterns may appear to change very slowly. To quantify this phenomenon in the simulations, we defined the spatial lag to be the maximum distance of a remnant population from the current range of suitable habitat. The maximum spatial lag in our simulations was 50, reflecting half the linear distance in the 100 x 100 arrays. Spatial lag was measured at the end of each simulation.

These simulations were all done with the initial proportion of habitat occupied by the community fixed at $p = 0.8$. This proportion established a landscape that was connected rather than fragmented (Gardner et al. 1987), permitting the community to migrate as a front. (Frontal migration could not occur if the potential habitat were fragmented). Rows of potential habitat added during the simulation also had 80% of their cells available for colonization. Simulation experiments were conducted with factorial combinations of the parameters: extinction ($e = 0.1$ and 0.5) corresponding to longevity values of $l = 10$ and 2 time steps, respectively; migration ($i = 0.1, 0.5$, and 1.0); and the interval of movement of potential habitat ($t = 1, 2$, and 10 time steps). For each combination of parameters, 10 simulations were conducted.

RESULTS

The proportion of the newly available habitat that was colonized by the community increased with increased

ORNL-DWG 89M-9755

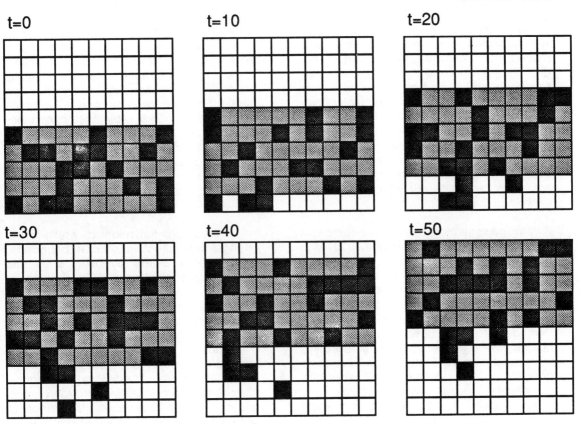

Figure 5. Illustration of the hypothetical movement of potential habitat and the response of a community. The grey area represents habitat that is suitable for colonization by a community type. Black cells are occupied by the community, and white areas are not suitable for colonization. At intervals of 10 time units, the potential habitat moves one row north. Remnant populations may remain behind, but colonization can only occur in the gray areas. The simulation concludes when the potential habitat reaches the last row of the map.

probabilities of migration (higher i) and slower rates of habitat movement (higher t) (Fig. 6). Provided that the probability of migration exceeded the probability of extinction, more new habitat was colonized at higher i values. The maximum distribution (p = 0.8) was approached when extinction was low and the probability of migration was greater than or equal to 0.5. For a given combination of e and i, more new habitat was colonized when the potential habitat moved more slowly (e.g., increasing t in Fig. 6). When the potential habitat moved very rapidly (e.g., t = 1), high values of i were necessary for the community to colonize only 40% of the available habitat. At high values of t and i and low values of e, the community could track the migration of suitable habitat. However, when extinction values were high, the proportion of the new habitat occupied by the community never exceeded 0.5 (Fig. 6).

The spatial lag in remnant populations was greatest when longevity of the community was high (longevity l = 10, corresponding to the lower extinction rate) and the potential habitat moved quickly (t = 1) (Fig. 7). At low extinction rates, the lag distance of the remnant populations decreased as the movement rate of the habitat slowed (higher values of t), reflecting increased opportunities for remnant populations to become extinct. When the community lifespan was short (l = 2, corresponding to the higher extinction probability), the spatial lag was very small (Fig. 7). For given values of t and e, an increased probability of migration generally increased the spatial lag because it allowed more community cells to become established.

The total amount of boundary in the landscape changed in response to the habitat displacement (Fig. 8). Initially, the total boundary length was approximately 4000, shown by the dashed line in Figure 8. The total amount of boundary was maintained or increased with particular combinations of e, i, and t (Fig. 8). With low extinction values, the amount of landscape boundary in the new habitat was similar to the initial habitat when the displacement rate was slow (higher t). This indicates a similar landscape structure and successful migration of the community. However, total landscape boundary decreased with low migration probability (i = 0.1) and rapid habitat displacement (low values of t) (Fig. 8). At higher extinction rates, the boundary increased by as much as 25%. As i increased, however, the total boundary in the landscape increased rapidly. At higher values of i and t, the amount of landscape boundary exceeded the initial conditions. This occurred because the proportion of the landscape occupied by the community approached 0.5 (Fig. 6), which is the value for which maximum boundaries are observed (Gardner et al. 1987).

DISCUSSION

Two potential means by which global change may affect landscape boundaries have been explored by using a "neutral model" approach (Caswell 1976, Gardner et al. 1987). Global climate change can alter the frequency and intensity of disturbances such as fires or pest outbreaks. In addition, global climate change can alter the spatial arrangement of species tolerance limits and change the location of boundaries between community types. Changing disturbance regimes were simulated in both fragmented and connected landscapes. The effects of potential habitat displacement were simulated only in landscapes with sufficient connectivity to allow migration to adjacent cells of suitable habitat.

The frequency and intensity of disturbances can be expected to change with the global climate, but the effects of changing disturbance regimes on landscape boundaries will be different in connected and fragmented landscapes. In landscapes that are connected (high p), an increase in disturbance intensity (probability of spreading) will cause landscape boundaries to decrease in total length. In landscapes that are fragmented (low p), an increase in

PROPORTION OF NEW HABITAT COLONIZED

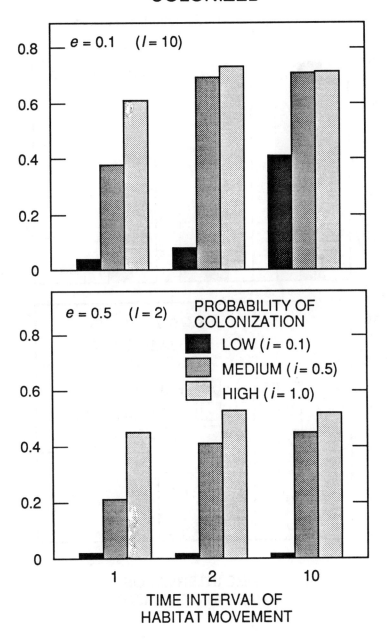

Figure 6. The proportion of newly available habitat that is successfully colonized by the community as a function of the rate of habitat movement and probability of dispersal, i, for (a) low probability of extinction and (b) high probability of extinction. When t = 1, habitat movement is rapid, and when t = 10, habitat movement is slow. The maximum possible community colonization is 0.8.

MAXIMUM LAG DISTANCE OF REMNANT COMMUNITIES

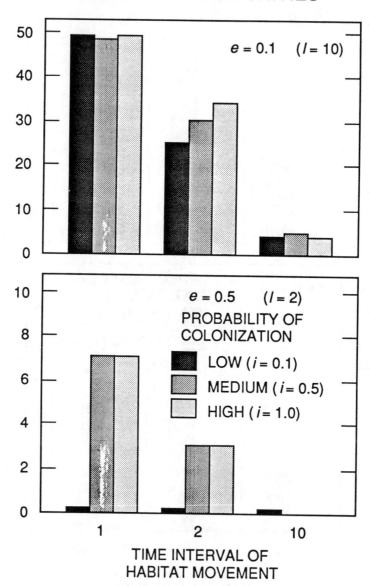

Figure 7. The maximum spatial lag distance of remnant communities from the suitable habitat as a function of the rate of habitat movement and probability of dispersal, i, for (a) low probability of extinction and (b) high probability of extinction. When t = 1, habitat movement is rapid, and when t = 10, habitat movement is slow. The maximum possible lag distance is 50, or half the size of the landscape maps.

LANDSCAPE BOUNDARIES

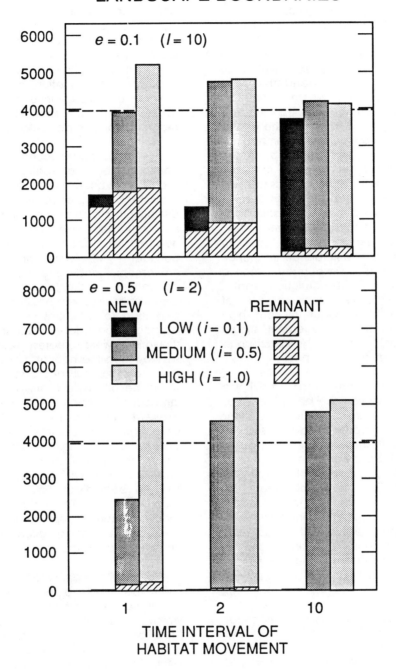

Figure 8. Landscape boundaries as a function of the rate of habitat movement and probability of dispersal, i, for (a) low probability of extinction and (b) high probability of extinction. Extinction probability is indicated by the variable, e, which is the inverse of longevity, l. The total boundary is the sum of boundaries in the newly available habitat and those of the remnant communities. When t = 1, habitat movement is rapid, and when t = 10, habitat movement is slow.

disturbance frequency (probability of disturbance initiation) will result in a decrease in landscape boundaries. The qualitative change in sensitivity to *i* and *f* occurs at the percolation threshold, p_c which is approximately 0.6 in random two-dimensional maps and will be different in more complex landscapes. This suggests different strategies for mitigating the effects of changing disturbance regimes in habitats that occur above and below p_c in the landscape. When a habitat is rare, the fragmented spatial distribution tends to inhibit disturbance spread, and the persistence of the habitat depends less on disturbance intensity than disturbance frequency. Therefore, management should focus on potential changes in disturbance frequency if the habitat is rare. In contrast, when a habitat is common, the contiguous spatial distribution enhances the spread of disturbances of low frequency but high intensity. Therefore, management should focus on potential changes in disturbance intensity if the habitat is common.

The conditions that maximize landscape boundaries can be determined, allowing general predictions about whether boundaries will increase or decrease with changing disturbance regimes. Habitat boundaries will be greatest in fragmented landscapes with disturbances of low intensity. Disturbance boundaries will be greatest in landscapes having intermediate values of *p* and disturbances of high intensity. As conditions depart from those that maximize either type of boundary (e.g., increasing *i* in fragmented landscapes), the simulations predict that the amount of boundary will decline.

Changes in the relative abundances of outer edges (perimeters of patches) and inner edges (gaps within larger habitat patches) may influence the distribution and abundance of wildlife species. Therefore, responses of inner and outer edges to global climate changes should be considered. The ratio of inner to outer edges was sensitive to the interaction between disturbance frequency and intensity in both fragmented and connected landscapes. Because these types of edge are qualitatively different (i.e., gaps or perimeters), changes in their relative abundance may have indirect effects on species or ecological processes. Predictions about the combination of *i* and *f* that maintain a particular inner/outer edge ratio might be useful in managing landscapes for the persistence of particular species.

Landscape boundaries are also controlled by the spatial distribution of species tolerance limits. Range limits are likely to be displaced from their current distributions in a changing climate, leading to new landscape patterns. The effects of habitat displacement on landscape boundaries will vary with the rate of movement and the probabilities of community extinction and migration. Climatically-induced habitat displacement may lead to time lags in the response of landscape boundaries. Spatial lags in the simulations were greatest when extinction was low and the habitat moved quickly. If this occurs in real landscapes, a time lag would be observed in the response of the landscape to habitat displacement. The present landscape could appear unchanged superficially, even though substantial changes occurred (e.g., community reestablishment is no longer possible). However, species responses are likely to be individualistic. By estimating parameters for this model for some key species, habitat displacement might be simulated by using predicted climatic changes through the next century.

Habitat displacement simulations might be used to guide mitigation by identifying the combinations of parameters which allow successful colonization and those for which community survival becomes unlikely. The simulated community could track the movement of habitat when the displacement rate was slow, the extinction probability was low, and the migration probability was at least moderate. Clearly, the more rapid the rate of habitat displacement, the more difficult it is for a community to migrate to its new range of potential habitat. Species with slow migration rates could be targeted for transplanting. An important topic for additional study is the influence of current landscape patterns on the ability of a species to successfully migrate to suitable habitat. Barriers such as urban centers or agricultural tracts may exert a dominant influence on species colonization.

CONCLUSIONS

The responses of landscape boundaries to global change undoubtedly will be complex. The simulations described in this paper predict interactions between landscape pattern, disturbance characteristics, and community migration characteristics. Because these predictions are quantitative, they provide testable hypotheses which can be evaluated for actual landscapes and used to examine the effects of different parameters. For example, a change in disturbance regimes can be simulated on real landscapes with disturbances of known frequencies and intensities. Similarly, habitat movement can be simulated for species with known lifespans and dispersal abilities and perhaps tested by modeling the range changes that accompanied past climatic change. A combination of theoretical and empirical work is necessary to better understand and predict the responses of landscape boundaries to global change.

ACKNOWLEDGEMENTS

Dr. Margaret B. Davis kindly provided unpublished material on species migration and time lags in response to climatic change. We also thank Yaffa L. Grossman, Marjorie M. Holland, Anthony W. King, Robert J. Naiman, Ronald P. Neilson, William H. Romme, and an anonymous reviewer for their thoughtful reviews of earlier drafts of the manuscript. Research funded by the Ecological Research Division, Office of Health and Environmental Research, U.S. Department of Energy, under Contract No. DE-AC05-84OR21400 with Martin Marietta Energy Systems, Inc. Publication No. 3628 of the Environmental Sciences Division, ORNL.

LITERATURE CITED

Baker, W. L. 1989. Landscape ecology and nature reserve design in the Boundary Waters Canoe Area, Minnesota, USA. Ecology **70**:23–35.

Brubaker, L. B. 1986. Responses of tree populations to climatic change. Vegetatio **67**:119–130.

Caswell, H. 1976. Community structure: a neutral model analysis. Ecological Monographs **46**:327–354.

Davis, M. B. 1984. Climatic instability, time lags, and community disequilibrium. Pages 269 – 284 in J. Diamond and T. J. Case, editors. Community ecology. Harper and Row, New York, New York, USA.

Davis, M. B. 1987. Invasions of forest communities during the Holocene: beech and hemlock in the Great Lakes Region. Pages 373–393 in A. J. Gray, M. J. Crawley, and P. J. Edwards, editors. Colonization, succession and stability. Blackwell Scientific Publications, Oxford, England.

Davis, M. B. 1989. Lags in vegetation response to greenhouse warming. Climatic Change **15**:75–82.

Davis, M. B. *In press.* Climatic change and the survival of forest species. *In* G. M. Woodwell, editor. The earth in transition: patterns and processes of biotic impoverishment. Cambridge University Press, Cambridge, England.

Davis, M. B., and D. B. Botkin. 1985. Sensitivity of cool-temperate forests and their fossil pollen record to rapid temperature change. Quaternary Research **23**:327–340.

Emanuel, W. R., H. H. Shugart, and M. P. Stevenson. 1985a. Climatic change and the broad-scale distribution of terrestrial ecosystem complexes. Climatic Change **7**:29–43.

Emanuel, W. R., H. H. Shugart, and M. P. Stevenson. 1985b. Response to comment: climatic change and the broad-scale distribution of terrestrial ecosystem complexes. Climatic Change **7**:457–460.

Forman, R. T. T., and M. Godron. 1986. Landscape ecology. John Wiley and Sons, New York, New York, USA.

Foster, D. R. 1988a. Disturbance history, community organization and vegetation dynamics of the old-growth Pisgah Forest, southwestern New Hampshire, USA. Journal of Ecology **76**:105–134.

Foster, D. R. 1988b. Species and stand response to catastrophic wind in central New England, USA. Journal of Ecology **76**:135–151.

Gardner, R. H., B. T. Milne, M. G. Turner, and R. V. O'Neill. 1987. Neutral models for the analysis of broad-scale landscape pattern. Landscape Ecology **1**:19–28.

Hansen, A. J., F. Di Castri, and R. J. Naiman. 1988. Ecotones: what and why? Biology International, Special Issue **17**:9–46.

Haynes, D. L. 1982. Effects of climate change on agricultural pests. Pages 1–42 *in* Environmental and societal consequences of a possible CO_2-induced climate change, Volume II, Part 10. DOE/EV/10019-10. National Technical Information Service, Springfield, Virginia, USA.

Knight, D. H. 1987. Parasites, lightning, and the vegetation mosaic in wilderness landscapes. Pages 59–83 *in* M. G. Turner, editor. Landscape heterogeneity and disturbance. Springer-Verlag, New York, New York, USA.

Mooney, H. A., and M. Godron, editors. 1983. Disturbance and ecosystems. Springer-Verlag, New York, New York, USA.

Pastor, J., and W. M. Post. 1988. Response of northern forests to CO_2-induced climate change. Nature **344**:55–58.

Pennington, W. 1986. Lags in adjustment of vegetation to climate caused by the pace of soil development: evidence from Britain. Vegetatio **67**:105–118.

Peters, R. L., and J. D. S. Darling. 1985. The greenhouse effect and nature reserves. BioScience **35**:707–717.

Pickett, S. T. A., and P. S. White, editors. 1985. The ecology of natural disturbance and patch dynamics. Academic Press, New York, New York, USA.

Risser, P. G., J. R. Karr, and R. T. T. Forman. 1984. Landscape ecology: directions and approaches. Illinois Natural History Survey, Special Publication Number 2. Champaign, Illinois, USA.

Romme, W. H. 1982. Fire and landscape diversity in subalpine forests of Yellowstone National Park. Ecological Monographs **52**:199–221.

Romme, W. H., and D. H. Knight. 1982. Landscape diversity: the concept applied to Yellowstone Park. BioScience **32**:664–70.

Runkle, J. R. 1985. Disturbance regimes in temperate forests. Pages 17–34 *in* S. T. A. Pickett and P. S. White, editors. The ecology of natural disturbance and patch dynamics. Academic Press, New York, New York, USA.

Rykiel, E. J. 1985. Towards a definition of ecological disturbance. Australian Journal of Ecology **10**:361–365.

Sandenburgh, R., C. Taylor, and J. S. Hoffman. 1987. Rising carbon dioxide, climate change, and forest management: an overview. Pages 113–121 *in* W. E.

Shands and J. S. Hoffman, editors. The greenhouse effect, climate change, and United States forests. The Conservation Foundation, Washington, D. C., USA.

Sellers, R. E., and D. G. Despain. 1976. Fire management in Yellowstone National Park. Pages 99–113 in Proceedings of the Tall Timbers Fire Ecology Conference, Number 14. Tall Timbers Research Station, Tallahassee, Florida, USA.

Solomon, A. M. 1986. Transient response of forests to CO_2-induced climate change: simulation modeling experiments in eastern North America. Oecologia 68:567–579.

Solomon, A. M., and T. Webb, III. 1985. Computer-aided reconstruction of late Quaternary landscape dynamics. Annual Review of Ecology and Systematics 16:63–84.

Stauffer, D. 1985. Introduction to percolation theory. Taylor and Francis, London, England.

Turner, M. G., editor. 1987. Landscape heterogeneity and disturbance. Springer-Verlag, New York, New York, USA.

Turner, M. G. 1989. Landscape ecology: the effect of pattern on process. Annual Review of Ecology and Systematics 20:171–197.

Turner, M. G., R. H. Gardner, V. H. Dale, and R.V. O'Neill. 1989. Predicting the spread of disturbance across heterogeneous landscapes. Oikos 55:121–129.

Urban, D. L., R. V. O'Neill, and H. H. Shugart. 1987. Landscape ecology. BioScience 37:119–127.

Webb, T., III. 1986. Is vegetation in equilibrium with climate? How to interpret late-Quaternary pollen data. Vegetatio 67:75–91.

White, P. S. 1979. Pattern, process, and natural disturbance in vegetation. Botanical Review 45:229–299.

White, P. S., and S. T. A. Pickett. 1985. Natural disturbance and patch dynamics: an introduction. Pages 3–13 in S. T. A. Pickett and P. S. White, editors. The ecology of natural disturbance and patch dynamics. Academic Press, New York, New York, USA.

Wiens, J. A., C. S. Crawford, and J. R. Gosz. 1985. Boundary dynamics: a conceptual framework for studying landscape ecosystems. Oikos 45:421–427.

SIMULATION OF THE SCALE-DEPENDENT EFFECTS OF LANDSCAPE BOUNDARIES ON SPECIES PERSISTENCE AND DISPERSAL

ROBERT H. GARDNER, MONICA G. TURNER, ROBERT V. O'NEILL AND SANDRA LAVOREL. Environmental Sciences Division, Oak Ridge National Laboratory, Oak Ridge, Tennessee 37831-6036, USA; and Centre d'Ecologie Fonctionnelle et Evolutive, Centre National de la Recherche Scientifique, Montpellier, France.

Abstract. The relationships between life history characteristics and broad-scale patterns of species abundance were investigated with a model that simulates the dispersal of populations through heterogeneous landscapes. Movement was simulated on randomly generated landscapes and on forested landscapes digitized from aerial photographs. Simulation results indicated that population abundances will change suddenly near the critical threshold in habitat connectivity as predicted from percolation theory. The existence of critical thresholds is important for many management and conservation issues, but these thresholds suggest that data are required at spatial scales that are specific to the dispersal characteristics of the simulated population.

Key words: landscape boundaries, species dispersal, habitat connectivity, life history effects, broad scale predictions.

INTRODUCTION

The problem of predicting the effects of extensive and rapid changes in land use patterns on the distribution and abundance of organisms continues to be an important ecological challenge (e.g., Burgess and Sharpe 1981, Forman and Godron 1981, Risser et al. 1984, Wiens et al. 1985b, Naiman et al. 1988). Many of the practical and theoretical problems associated with prediction might be resolved if the appropriate scales of interaction between pattern and process were understood (Wiens et al 1985a, Brown and Allen 1989, Carlile et al. 1989, Milne et al 1989). For example, measurements of landscape pattern depend on the scale at which the data are taken (e.g., Gardner et al. 1987, Meentemeyer and Box 1987, Urban et al. 1987, O'Neill et al. 1988). Theoretical studies based on fractal geometry (Mandelbrot 1967) show that measured boundary length changes as the scales of resolution change. Although simulation with a variety of spatial models (e.g., Gardner et al. 1987, Turner et al. 1989) can relate the number, size, and shape of landscape patches to the grain and extent of the map, this information is not sufficient to set the appropriate scales for predicting resource utilization by populations in heterogeneous environments.

Because species respond to environmental conditions at specific spatial and temporal scales (Swihart et al. 1988, Milne et al. 1989), it appears that an organism-centered view (sensu Wiens 1985) is essential for relating changes in landscape patterns to processes that affect the distribution and abundance of species. Previous applications of models based on percolation theory have shown that the spatial patterning of resources constrains the potential movement of populations across the landscape (O'Neill et al. 1988), with a sparse or patchy distribution of resources requiring organisms to operate at broader spatial scales.

This paper examines the proposition that the scale at which species are able to utilize landscape resources will be sensitive to the distance an organism (or propagule) must travel to colonize another site. The model we developed examines the effect of different combinations of population dispersal characteristics and the patterns of habitat boundaries on the predicted patterns of population growth, abundance, and habitat utilization. The results suggest that the mean and shape of the dispersal function (i.e., the probability of long distance movement) are critical determinants of the scale at which both data and model are required to predict population utilization in a patchy environment.

METHODS

A heterogenous distribution of habitats and their boundaries can be created by methods derived from percolation theory (Gardner et al. 1987) by randomly generating a map of m by m sites, with the habitat type of the m^2 sites selected with a probability of p. For sufficiently large maps an average of pm^2 sites will be designated as suitable habitat (e.g., a forested site) while $(1 - p)m^2$ sites will be designated as other (e.g., nonforested) sites. When the length, l, of a single site is specified (e.g., 100m for a 1-ha resolution), then the total area of the map is equal to $(lm)^2$. The scale of the map is defined by these two parameters: l is the resolution or grain size and $(lm)^2$ is the area or extent of the map.

Percolation theory defines a patch as a group of similar sites with at least one common edge along the vertical or horizontal direction. The number, size, and shape of the random patches are similar from map to map but will change as a function of p. Rapid changes in the size and shape of patches occur near the critical probability, p_c, when the largest patch just manages to extend from one edge of the map to the other. The value of p_c for extremely large maps has been empirically determined to be 0.5928 (Stauffer 1985). The relationship between patch size and boundary shape is also affected by p (Stauffer 1985): boundaries are short and straight when $p < p_c$, but longer and more convoluted when $p \geq p_c$.

The movement of populations across a heterogenous landscape can be simulated by two types of "rules" derived from

percolation theory: bond percolation "rules" simulate random, independent movement to individual nearest-neighbor sites; while site percolation "rules" simulate random, simultaneous movement to all nearest-neighbor sites (see von Niessen and Blumen 1988 for further description of bond and site percolation). When the probability of movement to new sites, i, is equal to 1.0, the movement patterns generated by bond or site percolation are equivalent and all sites within a patch will be colonized by the dispersing population. Differences between bond and site percolation become more apparent as the value of i decreases (see the Results section). Because habitat patches are highly fragmented when $p < p_c$, and because movement rules based on bond or site percolation only allow movement within a single patch, dispersal rules based on bond or site percolation methods will not allow the spread of a population across the map when $p < p_c$.

In addition to a model that simulates the spread of a population in a heterogenous landscape with dispersal rules based either on bond or site percolation methods, we have developed a model that simulates the spread of a population based on sets of functions which generate dispersal distances from continuous frequency distributions. The generation of dispersal distances from continuous frequency distributions relaxes the assumption of impermeable landscape boundaries and allows populations to colonize adjacent habitat patches. Continuous dispersal distances are generated by (1) selecting a random angle over the interval from 0^0 to 360^0; (2) selecting a random distance from some prespecified frequency distribution (uniform, triangular, or exponential); and (3) converting the angles and distances to appropriate map coordinates. The three types of continuous frequency distributions all produce circular patterns of dispersal but with different central tendencies. Although one might imagine that different dispersal functions can be used to represent different species characteristics (see Hanson, et al. 1989 for an example of this approach), the results reported are limited to the determination of the sensitivity of broad-scale predictions to potential differences in species dispersal characteristics.

The choice of parameters for the continuous dispersal functions affects both the shape and the distance that the organisms move from the parent population. If the maximum possible dispersal distance is less than $\frac{1}{2}l$, the population will not disperse beyond its current site. When the maximum possible dispersal distance equals l, the pattern of movement is similar to bond percolation. When the maximum possible dispersal distance is greater than l, then the habitat boundary is "permeable" and the population can disperse beyond the edge of the current habitat patch.

The model also considers other life history attributes, including the number of viable seeds, v, per generation per site; the probability of local population extinction, e (expected residence time is $1/e$); and the probability of site alteration, h, after local extinction. The parameters selected for the simulations reported here are analogous to populations of annual plants with a mean life expectancy of 10 generations (e = 0.1), whose seeds (v = 10) disperse a mean distance from the parent site. Because the site is not damaged by the population (h = 0.0), reinvasion after local extinction is possible. The parameters v, h, and e were held constant for all simulations.

Replicate simulations (N = 10) were performed on a series of random maps (m = 100 and l = 1) with different values of p. Landscape data on the distribution of forested sites in 9 counties in Georgia (Turner and Ruscher 1988) were used as input to the model to compare simulations with actual habitat patterns against simulations based on randomly generated maps. Aerial photographs for the 9 counties at 3 time periods were used to develop 27 different landscape maps. The nine counties included three from the piedmont, and two each in the mountains, upper coastal plain, and lower coastal plain. The photographically

interpreted information was assembled into 100 X 100 grids (each cell = 1 ha) with forested sites considered as the habitat of interest (see Turner and Ruscher 1988 and Turner 1990 for additional details).

RESULTS

Figure 1 compares the dispersal distances generated from the three continuous frequency distributions when the mean distance for all three distributions is equal to 1.0. The dispersal distances for the uniform distribution range from a minimum of 0.0 to a maximum of 2.0; the triangular distribution results in a minimum and mode of 0.0 and a maximum of 3.0; and 10,000 samples drawn from the exponential distribution produce a maximum distance of 10.0. The differences in the shapes of the distributions (Fig. 1) result in median values for the triangular (0.9) and exponential distribution (0.7) that are less than the uniform (1.0). However, the upper percentiles of the triangular and exponential distributions are 11% and 14% greater, respectively, than the maximum of 2.0 for the uniform distribution. Although the expected values for all three distributions are identical, differences in the shapes of the dispersal function create a higher probability of the triangular and exponential distribution, respectively, generating distances that result in species crossing habitat boundaries.

Figure 2 compares population growth through time, as measured by the number of map sites successfully populated, for four different sets of dispersal characteristics. The simulations were performed on a random landscape with p = 0.5 (50% of the sites are suitable habitat). The results show that when species movement is restricted to nearest-neighbor sites (bond percolation), less than 1% of the available habitat was utilized. When dispersal characteristics allow habitat boundaries to be readily crossed, then the rate at which a species spreads across the landscape increases for the uniform, triangular, and exponential distributions, respectively.

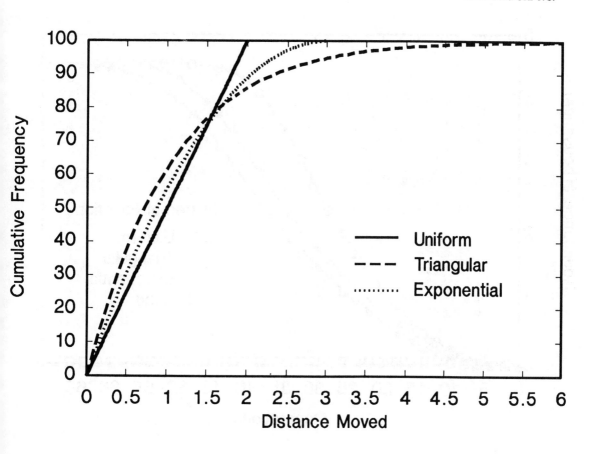

Figure 1. Cumulative frequency distributions for dispersal distances generated from the uniform, triangular, or exponential distributions. All three distributions have the same mean dispersal distance (mean = 1.0).

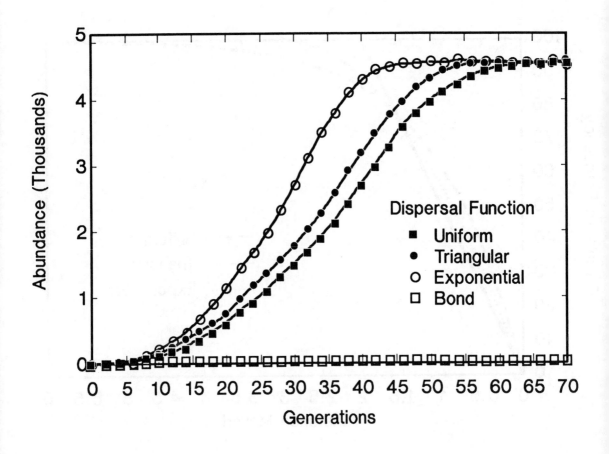

Figure 2. Comparison of the abundance through time of populations with different dispersal characteristics on a random landscape. Map dimensions are 100 rows by 100 columns with *p*, the fraction of sites that are suitable habitat, equal to 0.5 (5,000 total suitable sites per map).

If we assume that all available habitat sites can be reached by the simulated species, then the asymptotic population levels will be $pm^2(1.0 - e)$, where p is the fraction of suitable sites, m is the size of the grid, and e is the site-specific probability of population extinction. The asymptotic population level for the results illustrated in Fig. 2 is $0.5 \times 100^2 \times [1.0 -0.1] = 4500$ populated map sites. Because of differences in the maximum possible dispersal distances, the asymptotic levels were achieved at generations 46, 56, and 62 for the exponential, triangular, and uniform distributions, respectively.

A percolating map is defined as any map with a single, continuous habitat patch that spans the map. Thus, a map that "percolates" will allow a population that is restricted to nearest-neighbor movement (bond or site percolation with i = 1.0) to successfully spread across the map. The percolation characteristics of the 27 maps of Georgia landscapes were examined, and the results are illustrated in Figure 3. Although random maps show a sudden shift from nonpercolation below p_c to percolation above p_c, the Georgia landscapes show a broader range of values (p = 0.42 to 0.7) at which percolation may or may not occur. The two forested landscapes that determine the range of percolation values are illustrated in Figure 4. The photographs of Walker County taken in 1938 illustrate the ridge and valley system of this northwestern Georgia county. Because forests dominate the ridge tops (Fig. 4a), a continuous forest spans the map from north to south at p (the fraction of the grid points on the map occupied by forest) of only 0.43. The Herd County landscape, photographed in 1978, is dominated by forest habitat (p = 0.68). However, 33% of this Piedmont landscape is farmland or abandoned land concentrated in the upper third of the map, effectively disconnecting the northern forested areas from the southern portion of the map. Although the effects of topography, climate, and land use can be expected to cause differences from the expected patterns produced by random maps, only 2 of the 27

landscapes demonstrated percolation characteristics different from random (Fig. 3).

The similarities and differences between random and real maps are also illustrated by comparing the amount of habitat edge and the number of habitat patches on random and actual landscape maps (Fig. 5). The amount of edge on randomly generated maps is extremely consistent, with the maximum number of edges occurring at p = 0.5 (Fig. 5a), while the 27 Georgia landscapes show much less edge, with no apparent peak values. Because it is theoretically possible to produce "structured" maps with more edge than can be randomly generated (see Franklin and Forman 1987), it is apparent that the abiotic, human, and historical factors which structure actual landscapes produce patches of forested land with less edge than expected (see also Krummel et al. 1987).

The comparison of the number of habitat patches on random maps with the 27 maps of Georgia forested landscapes is shown in Figure 5b. A striking feature of this figure reveals that the random and real landscape maps are quite similar at values of p above 0.6. Although it is a trivial observation that no difference will exist between a random and an actual map when p = 1.0, Figure 5b indicates that differences between actual and random landscape patterns are rather small for values of p above p_c.

The pattern of **habitat utilization** (i.e., the number of sites occupied by the species of interest) on random and real landscapes is illustrated in Figure 6. The effect of movement patterns restricted to habitat patches (impermeable boundaries simulated by bond percolation) prevents much of the habitat from being utilized in disconnected landscapes; hence a sharp change in abundance occurs at $p = p_c$ (p_c = 0.5928 on random landscapes). A greater fraction of the Georgia landscape can be utilized by populations that do not cross the forest habitat boundaries (bond percolation, Fig. 6a) because these landscapes tend to be

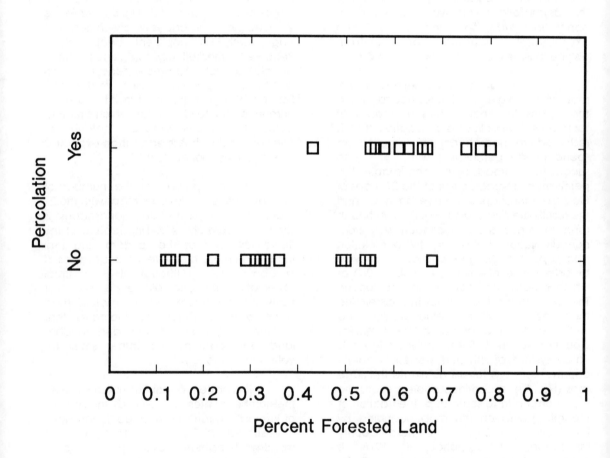

Figure 3. Percolation characteristics of forested areas taken from aerial photographs of nine counties in Georgia. A percolating map is one in which a single habitat patch spans the map.

Forested Land in Walker Co., GA 1938

Forested Land in Herd County, GA 1978

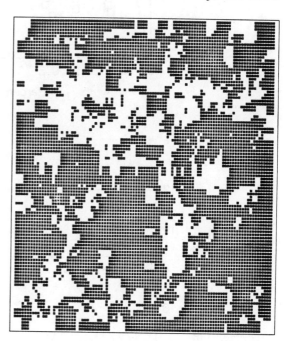

Figure 4. The dark areas represent forested regions of Herd County (Fig. 4a) and Walker County (Fig. 4b), Georgia. The data are based on aerial photos taken in 1978 for Herd County and 1938 for Walker County. Each grid point covers one hectare (see Turner and Ruscher 1988 and Turner 1990 for additional details). North is at the top of the figure.

83

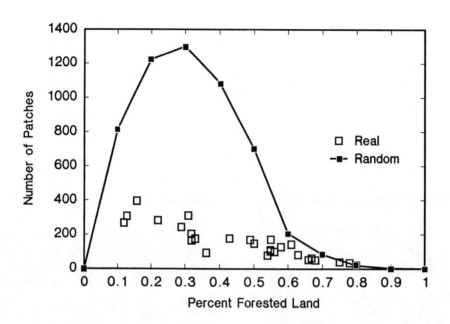

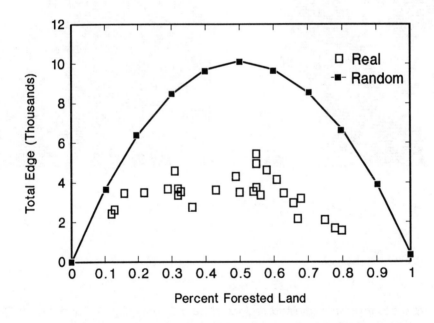

Figure 5. A comparison of habitat characteristics for randomly generated and actual landscape maps: a) Total habitat edge and b) Number of habitat patches.

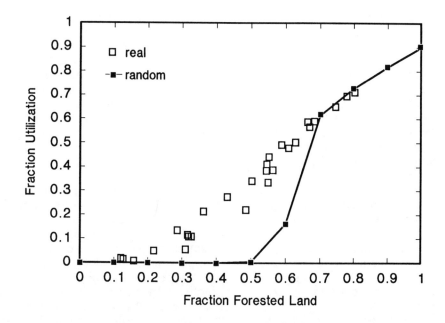

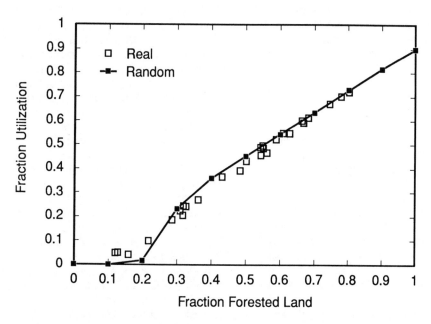

Figure 6. The fraction of habitat utilized on randomly generated and actual landscape maps for species dispersal characteristics which a) restrict dispersal to nearest-neighbor sites (bond percolation) or b) allow propagules to disperse a maximum of two-sites' distance from the parent population. Refer to the text for an explanation of algorithms used to generate dispersal distances.

connected over a broader range of values of p (Fig. 3) with fewer patches and fewer edges (Fig. 5). However, if the entire amount of available habitat were utilized, then a line through the points in Figure 5a could be drawn equal to $p *m^{2}* (1.0 - e)$, representing the asymptotic population levels which may be achieved under ideal conditions. The difference between any observation and this theoretical line is equal to the number of unpopulated but available habitat sites.

Movement by bond percolation (Fig. 6a) can be contrasted to movement patterns generated from the uniform distribution (Fig. 6b). Although the same maps and life history attributes were used for the two sets of simulations, Figure 6b shows that when dispersal distances are generated from a uniform distribution (minimum = 0.0, maximum = 2.0) the pattern of species abundance for random and real landscapes is essentially the same. Thus, when species movement patterns allow a small amount of permeability of the habitat edge, then the structure of the habitat (i.e., number and shape of suitable habitat patches) does not affect the fraction of the landscape that can be utilized.

CONCLUSIONS

The continuing loss of natural habitat and the associated decline in species abundance and diversity have made the design of natural reserves (i.e., the number, size, and spatial arrangement of critical habitat) an important conservation issue (Noss 1983, Noss and Harris 1986, Simberloff 1988). Arguments for many small isolated reserves are based on the observation that islands often have higher diversity per unit area than adjacent mainlands (Quinn and Hastings 1987). Studies of disturbance and recovery, however, argue for management policies that maintain habitat heterogeneity (Levin 1976a, Norse et al. 1986) and connectance between reserves (Fahrig and Merriam 1985). Other studies show that some species may be insensitive to both the size

and arrangement of critical resources (Fahrig and Paloheimo 1988). These different views of the effect of landscape design on biodiversity might be resolved by considering the scale at which organisms perceive and utilize natural resources (Levin 1976b, O'Neill et al. 1988). Because the scale at which species can distribute their progeny is fixed by their life history attributes, it is unlikely that fixed landscape patterns will be optimal for all species.

The simulations reported here allow the effect of landscape pattern and species dispersal characteristics to be systematically evaluated. The comparison of random maps to actual forested regions of nine counties in Georgia shows that the number (Fig. 5b), size, shape, and arrangement of habitat patches and their boundaries (Fig. 5a) are similar when the fraction of the forested landscape is greater than 0.6. The fact that actual landscapes show patterns similar to those predicted by studies of critical thresholds in percolation theory (Stauffer 1985, Gardner et al. 1987) confirms the usefulness of these concepts for studying the effects of landscape pattern on species abundances. When the fraction of forested landscape is below 0.6, then the effect of factors which "organize" actual landscape patterns (e.g., topography, land use history, human development, disturbance, etc.) can be expected to show "unique" results that are pattern (i.e., landscape) specific.

Simulations of the spread of populations across heterogeneous landscapes when the fraction of available habitat is above 0.6 show that results are not sensitive to species-specific differences in dispersal characteristics (Fig. 6a). Above 0.6, all landscapes are well connected and, no matter how far each species can disperse per generation, all populations will be able to reach a large fraction of the available habitat. Below 0.6, patches are highly fragmented, and simulations show that large differences in species abundances and habitat utilization are produced by small changes in the maximum possible dispersal distance. Species capable of moving greater distances are, therefore, not affected by landscape heterogeneity below the scale set by this dispersal parameter.

The landscape data necessary to identify critical thresholds for a variety of species--and thus produce an optimal design for the conservation of biodiversity--should not be gathered at fixed spatial scales. Rather, the spatial scales (i.e., the "grain" and extent of resources) should be determined for particular species and then the arrangement of critical resources evaluated at that scale. When the latter procedure is followed, it is possible to evaluate the effects of landscape pattern on the abundance and distribution of different species. Previous results have shown that above p_c habitat disturbance will increase the number of boundaries, but below p_c the effect will be to remove additional habitat and associated resources (Turner et al. 1989). Therefore, broad-scale differences in species abundance due to landscape structure are best studied on appropriately scaled maps with values of p ranging from 0.3 to 0.6.

Simulated population abundances of species which differ in dispersal characteristics show different rates of increase even though fecundity parameters are identical (Fig. 2). Because the converse is also true, that is, different landscape patterns result in different rates of increase for the same species, pattern and scale are critical in the management and design of natural resources. Further investigations will be needed to relate dormancy (Levin et al. 1984) and soil seed banks (Malanson 1984) to landscape patterns. Available field and experimental data lead us to expect that the presence of a seed bank will dramatically affect both the consequences of species dispersal characteristics and their response to disturbance.

Landscape studies of species abundance and diversity must be performed at scales appropriate to the species of interest. When this prerequisite is met, then new theory, methods, and empirical information can be used to relate pattern and process at broad spatial scales. The failure of many studies to first determine the appropriate scales for measurement and prediction may be responsible for the familiar complaint of lack of predictive theories in ecology (Pickett and White 1985).

ACKNOWLEDGEMENTS

The authors thank James Gosz, Yaffa L. Grossman, Marjorie M. Holland, Mari N. Jensen, Robert J. Naiman and an anonymous reviewer for comments on earlier drafts of the manuscript. Publication No. 3451, Environmental Sciences Division, Oak Ridge National Laboratory.

LITERATURE CITED

Brown, B. J., and T. F. H. Allen. 1989. The importance of scale evaluating herbivory impacts. Oikos **54**:189-194.

Burgess, R. L., and D.M. Sharpe, editors. 1981. Forest island dynamics in man-dominated landscapes. Springer-Verlag, New York, New York, USA.

Carlile, D. W., J. R. Skalski, J. E. Batker, J. M. Thomas, and V. I. Cullinan. 1989. Determination of ecological scale. Landscape Ecology **2**:203-213.

Fahrig, L., and G. Merriam. 1985. Habitat patch connectivity and population survival. Ecology **66**:1762-1768.

Fahrig, L., and J. Paloheimo. 1988. Effect of spatial arrangement of habitat patches on local population size. Ecology **69**:468-475.

Forman, R. T. T., and M. Godron. 1981. Patches and structural components for a landscape ecology. BioScience **31**:733-740.

Franklin, J. F., and R. T. T. Forman. 1987. Creating landscape patterns by forest cutting: ecological consequences and principles. Landscape Ecology **1**:5-18.

Gardner, R. H., B. T. Milne, M. G. Turner, and R. V. O'Neill. 1987. Neutral models for the analysis of broad-scale landscape pattern. Landscape Ecology **1**:19-28.

Hanson, J. S., G. P. Malanson, and M. P. Armstrong. 1989. Spatial constraints on the response of forest communities to climate change. Pages 1-23 in G. P. Malanson, editor. Natural areas facing climate change. SPB Academic Publishing bv, The Hague, The Netherlands.

Krummel, J. R., R. H. Gardner, G. Sugihara, and R. V. O'Neill. 1987. Landscape patterns in a disturbed environment. Oikos **48**:321-324.

Levin, S. A. 1976a. Population dynamic models in heterogeneous environments. Annual Reviews of Ecology and Systematics **7**:287-310.

Levin, S. A. 1976b. Spatial patterning and the structure of ecological communities. Lecture on Mathematics in the Life Sciences **8**:1-35.

Levin, S. A., D. Cohen, and A. Hastings. 1984. Dispersal strategies in patchy environments. Theoretical Population Biology **26**:165-191.

Malanson, G. P. 1984. Intensity as a third factor of disturbance regime and its effect on species diversity. Oikos **43**:411-413.

Mandelbrot, B. 1967. How long is the coast of Britain? Statistical self-similarity and fractional dimension. Science **156**:636-638.

Meentemeyer, V., and E. O. Box. 1987. Scale effects in landscape studies. Pages 15-34 in M. G. Turner, editor. Landscape heterogeneity and disturbance. Springer-Verlag, New York, New York, USA.

Milne, B. T., K. Johnston, and R. T. T. Forman. 1989. Scale-dependent proximity of wildlife habitat in a spatially-neutral Bayesian mode. Landscape Ecology **2**:101-110.

Naiman, R. J., M. M. Holland, H. Décamps, and P. G. Risser. 1988. A new UNESCO programme: research and management of land/inland water ecotones. Biology International, Special Issue **17**:107-136.

Norse, E. A., K. L. Rosenbaum, D. S. Wilcove, B. A. Wilcox, W. H. Romme, D. W. Johnston, and M. L. Stout. 1986. Conserving biological diversity in our national forests. Prepared by the Ecological Society of America for The Wilderness Society, Washington, D.C., USA.

Noss, R. F. 1983. A regional landscape approach to maintain diversity. BioScience **33**:700-706.

Noss, R. F., and L. D. Harris. 1986. Nodes, networks, and MUMs: preserving diversity at all scales. Environmental Management **10**:399-409.

O'Neill, R. V., B. T. Milne, M. G. Turner, and R. H. Gardner. 1988. Resource utilization scales and landscape pattern. Landscape Ecology **2**:63-69.

Pickett, S. T. A., and P. S. White. 1985. Patch dynamics: a synthesis. Pages 371-384 in S. T. A. Pickett and P. S. White, editors. The ecology of natural disturbance and patch dynamics. Academic Press, New York, New York, USA.

Quinn, J. F., and A. Hastings. 1987. Extinction in subdivided habitats. Conservation Biology **1**:198-208.

Risser, P. G., J. R. Karr, and R. T. T. Forman. 1984. Landscape ecology: directions and approaches. Illinois Natural History Survey, Special Publication Number 2. Champaign, Illinois, USA.

Simberloff, D. 1988. The contribution of population and community biology to conservation science. Annual Review of

Ecology and Systematics **19**:473–511.

Stauffer, D. 1985. Introduction to percolation theory. Taylor and Francis, London, England.

Swihart, R. K., N. A. Slade, and B. J. Bergstrom. 1988. Relating body size to the rate of home range use in mammals. Ecology **69**:393–399.

Turner, M. G. 1990. Landscape changes in nine rural counties in Georgia, USA. Photogrammetric Engineering and Remote Sensing **56**:379–386.

Turner, M. G., R. H. Gardner, V. H. Dale, and R. V. O'Neill. 1989. Predicting the spread of disturbance across heterogeneous landscapes. Oikos **55**:121–129.

Turner, M. G., and C. L. Ruscher. 1988. Changes in landscape patterns in Georgia, USA. Landscape Ecology **1**:241–251.

von Niessen, W., and A. Blumen. 1988. Dynamic simulation of forest fires. Canadian Journal of Forest Research **18**:805–812.

Wiens, J.A. 1985. Vertebrate responses to environmental patchiness in arid and semiarid ecosystems. Pages 169–193 *in* S. T. A. Pickett and P. S. White, editors. The ecology of natural disturbance and patch dynamics. Academic Press, New York, New York, USA.

Wiens, J. A., J. F. Addicott, T. J. Case, and J. Diamond. 1985*a*. Overview: the importance of spatial and temporal scale in ecological investigations. Pages 145–153 *in* J. Diamond and T. J. Case, editors. Community ecology. Harper and Row, New York, New York, USA.

Wiens, J. A., C. S. Crawford, and J. R. Gosz. 1985*b*. Boundary dynamics: a conceptual framework for studying landscape ecosystems. Oikos **45**:421–427.

Urban, D. L., R. V. O'Neill, and H. H. Shugart 1987. Landscape ecology. BioScience **37**:119–127.

HUMAN IMPACT ON THE FUNCTIONING OF LANDSCAPE BOUNDARIES

DAVID L. CORRELL. Smithsonian Environmental Research Center, Box 28, Edgewater, Maryland 21037, USA.

Abstract. Human management of the land frequently creates or sharpens landscape boundaries between natural and managed ecosystems. These boundaries often correspond approximately to natural landscape features, such as changes in slope, soil types, or riparian zones. Managed systems usually export very large fluxes of soils, nutrients, pesticides, and inorganic ions. Steep gradients of material concentrations and process rates develop across and adjacent to these boundaries. The resulting alteration in rates of processes in the natural ecosystem ultimately has secondary unanticipated effects upon soils, biota, and atmospheric exchange rates. In the below ground environment, pH, Eh, ionic composition, nutrient status, toxic metal levels, and organic matter pathways are altered. These impacts are especially important for remnant or relict natural ecosystems such as wetlands which are downslope from intensively managed ecosystems.

Key words: landscape boundary, managed landscapes, riparian forests, cropland, coastal plain.

INTRODUCTION

Both managed and unmanaged landscapes have internal boundaries. Some, such as those between water and land, between various plant and animal communities, or between emergent bedrock and surface soils, are apparent and clearly seen. Others, such as zones of differing pH, Eh, mineralogy, or salinity may be less apparent unless measured with arrays of instrumental sensors.

Few landscapes on this planet remain unmanaged. Human management of the land often creates new boundaries or sharpens previous ones. Boundaries between managed and unmanaged soils and plant communities are a common feature of most landscapes (Risser et al. 1984, Wiens et al. 1985, Naiman et al. 1988). The unmanaged areas of these landscapes become relict habitat patches or corridors of relatively natural habitat, often heavily stressed by their managed surroundings (stress in the sense of Barrett et al. 1976).

Less apparent, invisible boundaries between managed and unmanaged areas are also usually altered and sharpened.

These boundaries involve steep gradients in various parameters which are produced by both slow processes such as pedogenesis and by rapid processes such as dominant below ground electron transport reactions. Managed areas often export large fluxes of eroded soil, water, electron acceptors such as nitrate and sulfate, mineral nutrients, insecticides, herbicides and other pesticides, and toxic metals (Harper 1974).

Unmanaged relict natural habitat areas often react to the stress imposed by these large fluxes from adjacent managed areas with unbalanced high process rates which result in altered soils, biota, and rates of reactant emissions. These gas emissions are increasingly recognized as globally important fluxes (Mooney et al. 1987). Thus, one might argue that we should study newly created landscape boundaries as successional zones. For example, when humans remove an upland forest and plant row crops, a successional sequence is initiated leading to rapid alteration of below ground processes and the slow alteration of soils and plant and animal communities along the newly created boundary. If the land management practiced on the row crops is later altered, this will also be reflected in the dynamics of the boundary between managed and unmanaged areas.

In general, visually discernable boundaries have been better studied and environmental scientists are only beginning to characterize the invisible, process-related boundaries which are often the causal agents for some of the visible characteristics of these boundaries (Correll and Weller 1989, Pinay et al. 1989).

In the following sections I will: a) review some published data, b) present some relevant new data, and c) attempt to generalize and speculate about human impact on landscape boundary functions. I will focus on the specific case where uplands are managed in row crops and where the lowlands remain as streamside, deciduous hardwood forests.

METHODS

Site Description

The internal dynamics of eight riparian forests were studied on the inner coastal plain of Maryland (38° 53' N, 76° 35' W). The general site is within the Eocene Nanjemoy formation. The 16 ha primary stream drainage (watershed 109) has an average slope of 5.4%. The soils of the upper part are of the Collington series while those of the lower part are of the Westphalia series. All are fine, sandy loam of sedimentary origin. The most abundant clay and silt mineral is montmorillonite $[Al_2Si_4O_{10}(OH)_2]$ followed by illite (interlayered mica and montmorillonite) followed by kaolinite $(H_4Al_2Si_2O_9)$ (Correll et al. 1984). Bedrock is approximately 1,000 m below the surface but the watershed is perched on an impervious clay layer (the Marlboro Clay) which acts as an effective aquiclude. Surface soils in the cropland have an average pH of 5.6 and an organic matter content of 1.9%. Below the artificial soil plow horizon, pH and organic content drop rapidly (Correll 1983).

The riparian forest studied is the upland end-member of a wetland continuum including, in sequence, floodplain forests, forested swamps, and tidal marshes (Correll and Weller 1989). The riparian forest vegetation is dominated by sweetgum (*Liquidambar styraciflua*) and red maple (*Acer rubrum*) with tulip poplar (*Liriodendron tulipifera*) as an important constituent in some areas (Peterjohn and Correll 1984).

The sites of the small watershed-level comparative studies are all on the coastal plain portion of the Chesapeake Bay watershed. Each station is on a low order (first, second or third order) stream. Sites were selected partially on the basis of the characteristics of their riparian zone vegetation. They differ in the width, continuity, and type of vegetation found in their riparian zones (Table 1).

Sampling

Long-term watershed study sites are all equipped with sharp-crested vee-notch weirs. Stage height is recorded every 5 or 15 minutes depending on drainage basin size, and volume-integrated composite samples are taken automatically (Correll 1977, 1981). Sites selected for two year comparative studies are equipped with stilling wells and instrument enclosures. The hydrograph of the stream open-channel is monitored with a Druck pressure transducer. A Cambell Scientific CR-10 data logger/microcomputer system and a stream discharge rating curve are used to store discharge rate data and to control volume-integrated water samplers. Stream water samples are preserved and analyzed as described by Peterjohn and Correll (1984).

Many of the results reviewed here have been previously published and the methods by which they were obtained may be found in the citations. Measurements of nitrous oxide emission rates and concentrations in groundwater have not previously been reported. An area extending from 10 m within the cropland to 20 m inside the forest and extending 60 m along the cropland/forest boundary was marked with a 10x10 m grid. A stratified

Table 1. Summary of watershed characteristics.

DISCHARGE RESULTS **

Watershed Number	Area of Cropland (%)	Riparian Vegetation *	Nitrate	Total Ammonium	Total P	Atomic Ratio Total N/P
108	47	CWF	I	H	H	L
110	0	F	L	L	L	I
119	37	CWF	I	L	I	L
301	0	F	L	L	L	I
305	80	CNF	H	I	I	H
306	80	NH	H	H	H	I
307	0	F	L	L	L	H
308	70	WH	H	L	L	H

*C = continuous, N = narrow, W = wide, F= forest, H = herbaceous

**L = low, I = intermediate, H = high

random set of sampling locations was predetermined. Semi-closed chambers as described by Hutchinson and Mosier (1981) were used to determine N_2O emission rates from the soil surface. Each determination was made from a set of gas samples taken from a chamber's headspace at 0, 20, 40 and 60 minutes from time of chamber placement.

Sample Analyses

Gas samples were analyzed for N_2O by gas chromatography on a 3 m porapak-Q column at 34° with an Ni[63] electron capture detector. Shallow groundwaters (1-3 m) were also sampled from piezometers and analyzed for dissolved N_2O by static gas partitioning (Kaplan et al. 1978). Total dissolved aluminum was also measured on samples taken from piezometers. Samples were filtered through 0.4 μm nucleopore membranes which had been prewashed with acid, then distilled water. Samples were acidified to pH 2 immediately after filtration with ultrapure concentrated HNO_3. Total dissolved Al was determined with a Perkin/Elmer, model 5000, atomic absorption spectrometer on a temperature stabilized platform in a Zeeman graphite furnace in the presence of $Mg(NO_3)_2$.

RESULTS AND DISCUSSION

I will summarize what we have learned in the last 12 years about the functioning of one particular type of landscape boundary. This is a boundary between upland crops and first order stream riparian deciduous hardwood forest. This boundary is the most prevalent human-made landscape boundary in the Atlantic coastal plain geological province. We have been conducting a case study of a particular 6 ha riparian forest surrounded by 10 ha of upland fields in continuous corn production. The entire area was originally a continuous hardwood forest, but humans cleared the uplands in the mid-sixteenth century. Although the riparian zone has been selectively logged, it was allowed to

remain in forest because it was too wet to farm with conventional methods. Thus, for a long time period, humans have created and maintained this artificial landscape boundary.

Below ground process rates across the boundary

In the fields devoted to corn production approximately 140 of a total fertilizer load of approximately 170 Kg N/ha/yr was applied as reduced forms of nitrogen. The mineralization and nitrification of this fertilizer and parent soil organic nitrogen produce high concentrations of nitrate and hydronium ion, much of which is washed from the cropland into the riparian forest in shallow groundwaters (Peterjohn and Correll 1986, Correll et al. 1987). Over 70% of the hydronium ions exported from the cropland were due to nitrification rather than acidic atmospheric depositions (Correll et al. 1987). The ground waters entering the riparian forest from the cropland during two different years contained mean nitrate-N concentrations of 6.2 and 4.5 mg N/l and mean pH's of 4.6 and 4.5 respectively (Peterjohn and Correll 1986).

The impacts on below ground processes in the riparian forest include the consumption of significant amounts of forest-produced organic matter to serve as an electron donor and the consumption of all available dissolved oxygen in the ground-water, the depletion of most of the nitrate and some of the available sulfate as electron acceptors (Fig. 1a, b, Fig. 2; Correll and Weller 1989). Approximately 25% of the nitrate was assimilated and stored in accumulated woody biomass and most of the rest is believed to have been denitrified. Both processes utilize or neutralize hydronium ions (Correll and Weller 1989). Thus, groundwater leaving the riparian forest in the two study years had average nitrate-N concentrations of 0.8 and 0.9 mg /l and average pH's of 5.5 and 5.5 (Peterjohn and Correll 1986). Similar utilization of nitrate from groundwater has also been reported by other groups (Lowrance et al. 1984a,b,c,

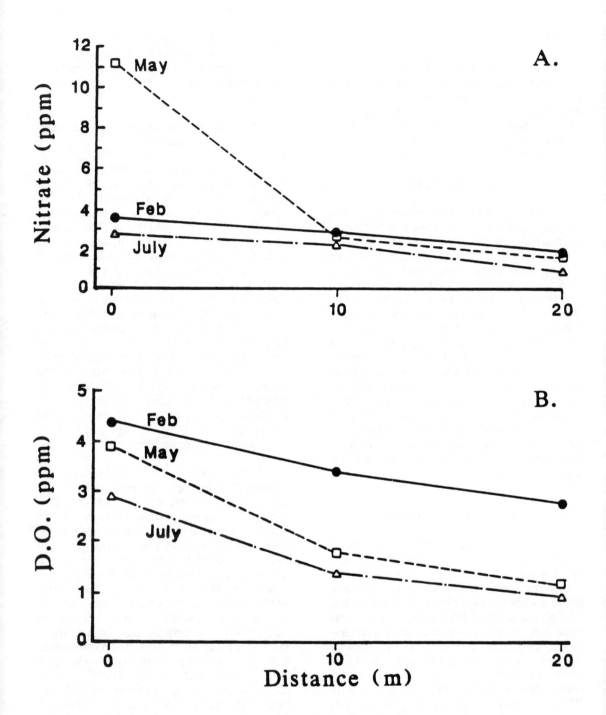

Figure 1. Changes in nitrate concentrations (Panel A) and dissolved oxygen (Panel B) in groundwater traversing from cropland to stream (in 1984) through riparian forest on Watershed 109 (from Correll and Weller 1989).

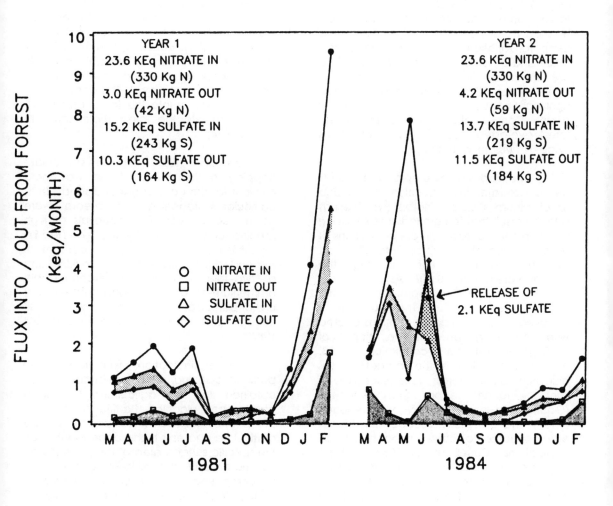

Figure 2. Time series of nitrate and sulfate fluxes through riparian forest in groundwater discharged from croplands on Watershed 109.

Jacobs and Gilliam 1985*a,b*, Cooper et al. 1986, Labroue and Pinay 1986, Pinay and Labroue 1986, Pinay and Décamps 1988).

Evidence that denitrification is a significant process and that it increased as groundwater moved from cropland to forest was obtained by measuring short-term N_2O emission rates from the soils with chambers (Fig. 3). Rates of N_2O emission increased rapidly across the cropland/riparian forest boundary. However, very high spatial and temporal variation in rates of emission made these and other such measurements of little use in calculation of annual rates of denitrification. Groundwater also contained significant concentrations of dissolved nitrous oxide. Concentrations 10 to 20 m into the forest averaged about 100 $\mu g/l$ of nitrous oxide-N (weight/volume).

The acidity of the cropland soil waters also accelerates the dissolution of clay and silt aluminum silicate minerals (weathering) with the result that the groundwaters leaving the cropland contain high concentrations of dissolved aluminum. Typical total dissolved aluminum concentrations leaving the cropland are about 400 $\mu g/l$. As these acidic groundwaters become neutralized in the riparian forest, much of the aluminum is removed. Aluminum concentrations decline with increasing pH with a slope of $-410 \mu g/l/pH$ unit and a correlation of 0.73. An open question is the impact of this aluminum on the forest trees, especially in light of recent reports from Germany (Godbold et al. 1988). Even so, enough acidity and aluminum escape into the first order streams to cause impacts on fish populations (Correll et al. 1987). These processes and their spatial pattern can be conceptualized as in Figure 4. Any cropland/riparian forest system might be expected to behave in a similar manner if the groundwater hydrology is analogous.

Surface process rates across the boundary

The most apparent human-induced flux from cropland to riparian forest is suspended soil particulates in storm induced overland flows, especially in the spring and summer. Mean suspended particulate concentrations were 8.5 and 11 g/l leaving the cropland in spring and summer, respectively. These were reduced to 1.4 and 1.0 g/l by the time overland flows penetrate in the forest 19 m and to 0.4 and 0.5 g/l after 50 m penetration into the forest (Peterjohn and Correll 1984). Most of the high nutrient concentrations in cropland overland flows were associated with the suspended particulates and were reduced commensurately. An exception was nitrate-N which had an annual average concentration in overland flow discharges from cropland of 4.4 mg/l. This was reduced to 1.8 mg/l 19 m into the forest and to 0.9 mg/l 50 m into the forest (Peterjohn and Correll 1984).

As one might expect, particulate trapping in the riparian forest was more efficient for the coarser silts and sand. Thus, particulates reaching the stream were dominated by very fine materials of high specific surface area. This was reflected by an increase in the organic-C content of suspended particulates from 1.5 to 8.2% and a two-fold increase of exchangeable ammonium-N and orthophosphate-P per unit of particulate mass during a 50 m transit through the forest (Peterjohn and Correll 1984).

Similar indications with respect to particulate sorting in the riparian forest were obtained for the agricultural herbicides atrazine and alachlor. Partitioning of these compounds between water and suspended particulate phases is highly dependent upon the specific surface area of the particulates. Kd, the dimensionless ratio of concentration of compound per kg of particulates to concentration of compound per kg of solvent, increased from about one in the surface soils of the cropland to annual averages in the primary stream of 114,000 and 148,000 respectively, for atrazine and alachlor (Correll et al. 1978).

Thus, this riparian forest has trapped much of the high load of suspended soil

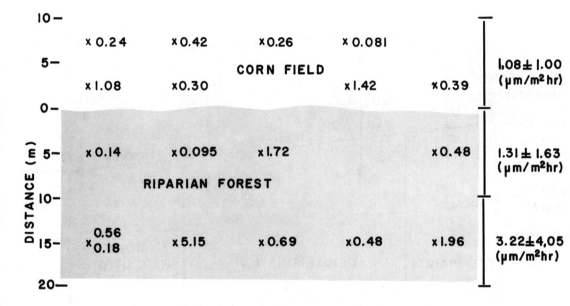

Figure 3. Spatial variation of emission rates for nitrous oxide from cornfield and riparian forest soils at various distances from the boundary between these ecosystems on Watershed 109. Values for individual rate measurements were obtained at sites marked with X's. Values at the right are means and standard deviations for each zone.

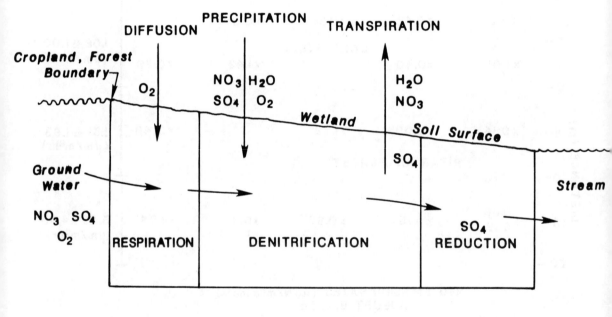

Figure 4. Conceptual cross-section of below ground processes affecting chemical fluxes in riparian forest (from Correll and Weller 1989).

particulates and particulate nutrients carried into the forest from croplands as storm-induced overland flow. It also removed most of the dissolved nitrate from the overland flows. The forest was also exposed to significant herbicide concentrations both as dissolved and suspended particulate bound fractions (Correll et al. 1978).

Another functional change across the cropland/riparian forest boundary is the rate of evapotranspiration (ET). A chloride balance was used to aid in the determination of annual ET estimates for the cropland and forest for a two year period. The cropland ET rates were 60.8 and 64.8 cm/yr, for the two years. Forest ET rates were 107 and 118 cm/yr (Peterjohn and Correll 1986). The riparian forest was able to sustain these high evapotranspiration rates in part because the cropland ET was low most of the year, allowing more groundwater movement into the riparian forest than would be the case if the uplands were mature hardwood forest.

Interwatershed comparisons

Mean nutrient concentrations in discharges from a series of watersheds were compared for the months of April, May and June, 1989. This was a period of unusually high precipitation for this region (511 mm versus a 160 year average of 281 mm; Higman and Correll 1982). Results for eight watersheds are summarized in Figure 5 and Table 1. Watersheds 110 and 301 are completely forested and are on the inner coastal plain, while watershed 307 is also completely forested and is on the outer coastal plain. Watersheds 119 and 108 are on the inner coastal plain and have continuous, fairly wide riparian forest. They differ in the proportion of row-cropped upland (119 = 37%, 108 = 47%, Correll 1977). Watersheds 305 and 306 are in the middle coastal plain and both have approximately 80% cropland. In watershed 305, the other 20% is primarily a narrow but continuous riparian forest, while on watershed 306 there is a similar proportion of forest, but it is mostly upland forest and most

of the stream channels have little or no riparian vegetation. Watershed 308 is on the outer coastal plain, has about 70% cropland, and has fairly continuous riparian vegetation. However, this riparian vegetation is predominantly herbaceous with few trees.

Nitrate discharges (Fig. 5) were very low from the three forested watersheds (110, 301, 307), intermediate from 119 and 108, high from 305 and 306, and highest from 308. Ammonium discharges were fairly low from all but 108 and 306. Total nitrogen discharges, which include organic-N, were lowest for 301 and 307, intermediate for 110 and 119 and much higher for the rest. Inorganic phosphate discharges were much as expected from the land use and geology with the exception of watershed 308. Watersheds with more cropland or less riparian vegetation had higher discharges. Inner coastal plain discharges were highest and outer coastal plain discharges were lowest. Watershed 308 phosphate discharges were unusually low and this may be due to the dense herbaceous riparian vegetation on this watershed, which seems to be an effective trap for eroded particulates with their high phosphorus content. However, watershed 308 nitrate discharges were highest, indicating ineffective nitrate interception by this herbaceous riparian zone. The importance of the type and extent of riparian vegetation is clearly reflected in the atomic ratio of total-N to total-P in the discharges. These ranged from 4.0 for watershed 108 to 44 for watershed 308.

Time courses of discharges for total-N and P from various watersheds also present some clues to the nutrient buffering of various patterns and types of riparian vegetation. Thus completely forested watershed 110's N and P discharges (Fig. 6) track each other closely during and between storm events. Watershed 119 has a lower N/P ratio, but N and P discharges have similar trajectories (Fig. 7). Watershed 108, a more highly agricultural basin, has a lower overall and much more variable N/P ratio in

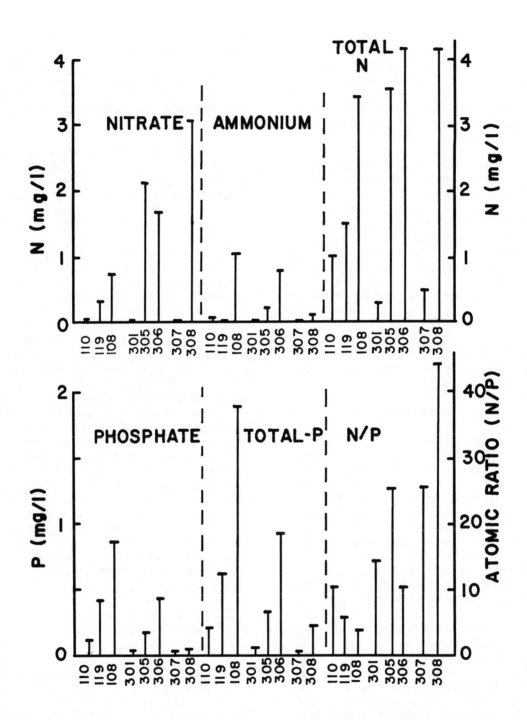

Figure 5. Comparative volume integrated average nutrient concentrations for various research watersheds for April, May and June of 1989. Characteristics of each watershed are described in the text and in Table 1.

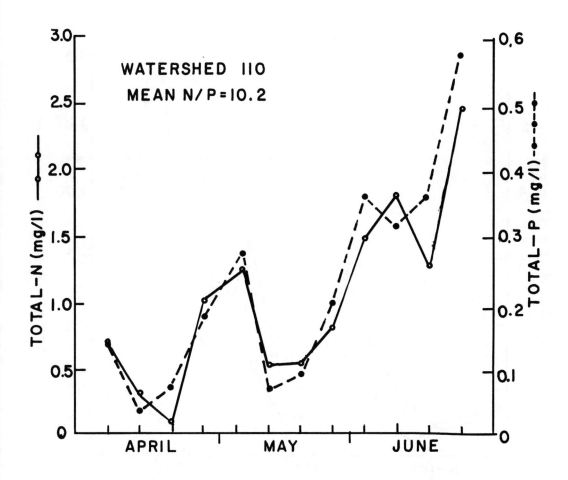

Figure 6. Time series in 1989 of weekly volume-integrated nutrient concentration data for a first order completely forested watershed on the inner coastal plain. The average atomic ratio of N to P is noted.

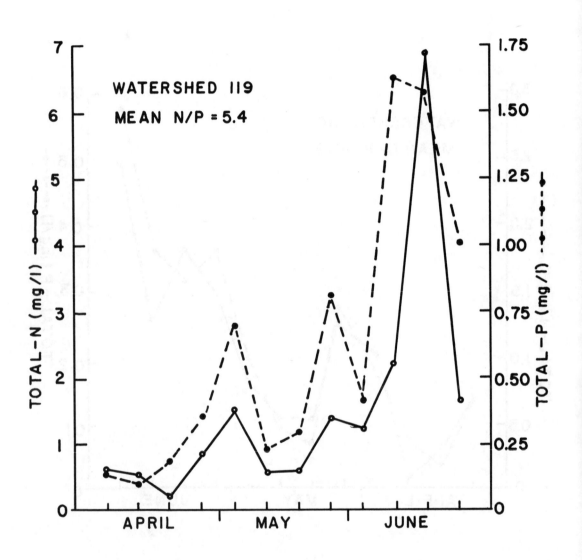

Figure 7. Time series in 1989 of weekly volume-integrated nutrient concentration data for a third order mixed land use watershed on the inner coastal plain with 37% upland cropland. The average atomic ratio of N to P is noted.

discharges (Fig. 8). Only during the last storm event did the N/P ratio decline dramatically. Watersheds 305 (Fig. 9) and 306 (Fig. 10), which are both in the middle coastal plain and both of which have about 80% cropland, behave quite differently from each other. Watershed 306, with little riparian forest, releases high phosphorus discharges during each storm event and has a lower overall N/P atomic ratio than watershed 305, which has continuous riparian forest. Watershed 308 (Fig. 11) had extremely high atomic N/P ratios (on the order of 100) except during the largest storm at the end of this time series, when the concentration of phosphorus increased quite significantly.

We know very little about the comparative functional effectiveness of: a) deciduous hardwood forest versus coniferous forest, b) old forest versus young forest and brush, or c) forest versus grass or herbaceous vegetation. Likewise, we know very little about the importance of groundwater trajectories and soil types. Finally, we do not have a sufficient understanding of processes and the controls over these processes to build adequate simulation models of the responses of riparian vegetation to material fluxes from upslope agricultural lands. Therefore, it is presently impossible to accurately predict how land use changes and climatic changes will affect these systems in the future.

CONCLUSIONS AND RECOMMENDATIONS

One could hypothesize that stream riparian zones could be designed and managed to attain dramatically lowered nutrient discharges from upland areas of row crops. One could maintain mixed hardwood forest along the stream banks, where the soils are water logged and conditions are optimal for nitrate interception. A second zone of herbaceous vegetation could be established contiguous to the forest and below the cropland for the purpose of intercepting eroded soil particulates with their high content of phosphorus and reduced nitrogen. This herbaceous zone would also help prevent gully formation and would help prevent concentrated storm flows from surging through the riparian forest before adequate time has elapsed for nutrient processing.

In regions which were naturally completely forested, clearing and fragmentation of the forest by humans has often resulted in seriously reduced receiving water quality. A few studies have demonstrated the important buffering functions of remnant riparian forests. However, there is a great need for further research on the functions of riparian vegetation in these agricultural landscapes.

ACKNOWLEDGEMENTS

The author thanks M.M. Holland, M.N. Jensen, M.G. Turner, R.J. Naiman, P.G. Risser, and an anonymous reviewer for reviews of earlier drafts of this manuscript. This research has been supported, in part, by the Smithsonian Institution's Environmental Sciences Program and by NSF grants CEE-8219615, BSR 79-11563, BSR 82-07212, BSR 83-16948 and BSR 86-15902.

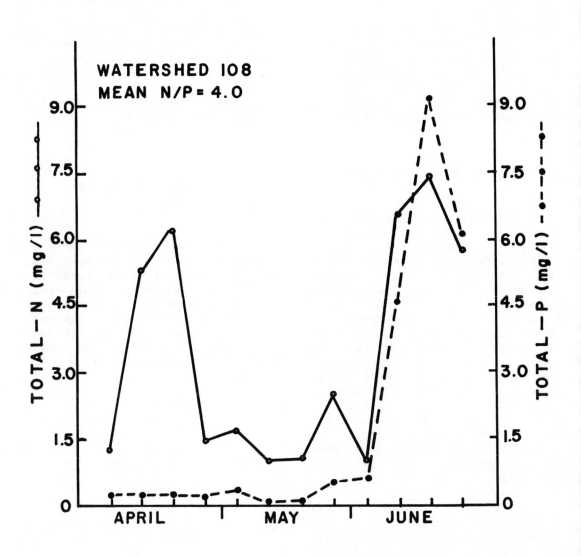

Figure 8. Time series in 1989 of weekly volume-integrated nutrient concentration data for a second order mixed land use watershed on the inner coastal plain with 47% upland cropland. The average atomic ratio of N to P is noted.

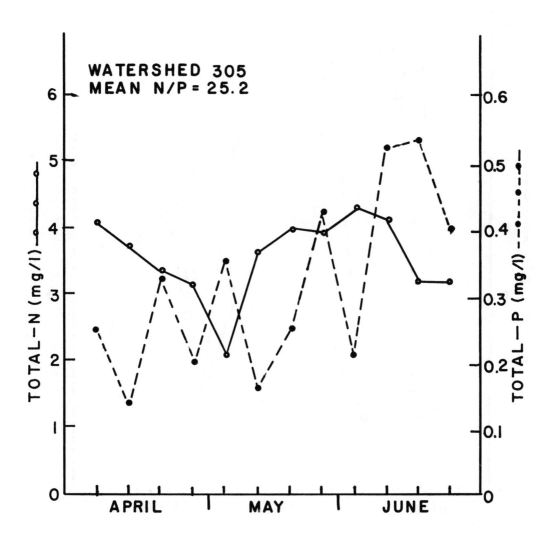

Figure 9. Time series in 1989 of weekly volume-integrated nutrient concentration data for a third order middle coastal plain watershed with about 80% cropland and continuous but narrow riparian forest. The average atomic ratio of N to P is noted.

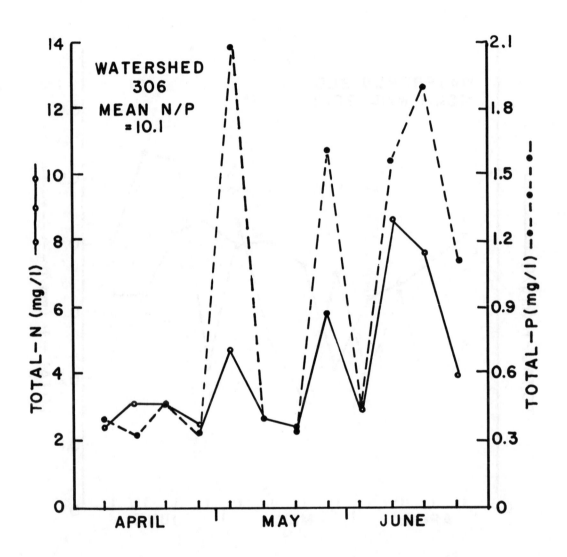

Figure 10. Time series in 1989 of weekly volume-integrated nutrient concentration data for a third order middle coastal plain watershed with about 80% cropland and little riparian forest. The average atomic ratio of N to P is noted.

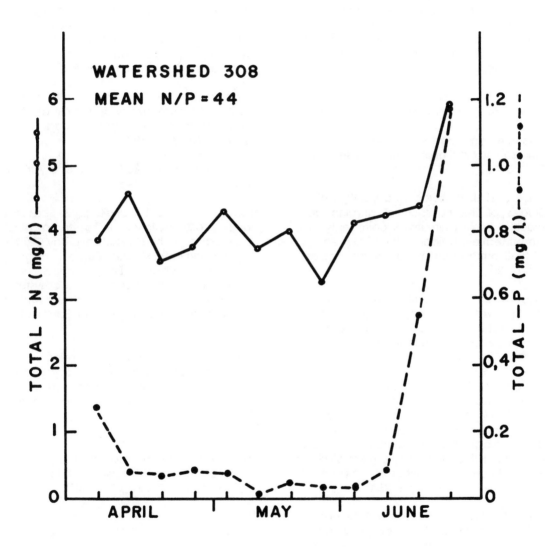

Figure 11. Time series in 1989 of weekly volume-integrated nutrient concentration data for a second order outer coastal plain watershed with about 70% cropland and continuous herbaceous riparian vegetation. The average atomic ratio of N to P is noted.

LITERATURE CITED

Barrett, G. W., G. M. Van Dyne, and E. P. Odum. 1976. Stress Ecology. BioScience 26:192-194.

Cooper, J. R., J. W. Gilliam, and T. C. Jacobs. 1986. Riparian areas as a control of nonpoint pollutants. Pages 166-192 in D. L. Correll, editor. Watershed research perspectives. Smithsonian Institution Press, Washington, D.C., USA.

Correll, D. L. 1977. An overview of the Rhode River watershed program. Pages 106-124 in D. L. Correll, editor. Watershed research in eastern North America. Smithsonian Institution Press, Washington, D.C., USA.

Correll, D. L., J. W. Pierce, and T. L. Wu. 1978. Studies of the transport of atrazine and alachlor from minimum till corn fields into Chesapeake Bay tidal waters. Proceedings of the Northeastern Weed Science Society 32:21-32.

Correll, D. L. 1981. Nutrient mass balances for the watershed, headwaters intertidal zone, and basin of the Rhode River estuary. Limnology and Oceanography 26:1142-1149.

Correll, D. L. 1983. N and P in soils and runoff of three coastal plain land uses. Pages 207-224 in R. Lowrance, R. Todd, L. Asmussen, and R. Leonard, editors. Nutrient cycling in agricultural ecosystems. University of Georgia Agricultural Experiment Station, Special Publication Number 23. University of Georgia, Athens, Georgia, USA.

Correll, D. L., N. M. Goff, and W. T. Peterjohn. 1984. Ion balances between precipitation inputs and Rhode River watershed discharges. Pages 77-111 in O. P. Bricker, editor. Geological aspects of acid deposition. Butterworth, Boston, Massachusetts, USA.

Correll, D. L., J. J. Miklas, A. H. Hines, and J. J. Schafer. 1987. Chemical and biological trends associated with acidic atmospheric deposition in the Rhode River watershed and estuary. Water, Air and Soil Pollution 35:63-86.

Correll, D. L., and D. E. Weller. 1989. Factors limiting processes in freshwater wetlands: an agricultural primary stream riparian forest. Pages 9-23 in R. R. Sharitz and J.W. Gibbons, editors. Freshwater wetlands and wildlife. Savannah River Ecology Laboratory, Aiken, South Carolina, USA.

Godbold, D. L., E. Fritz, and A. Huttermann. 1988. Aluminum toxicity and forest decline. Proceedings of the National Academy of Science 85:3888-3892.

Harper, J. L. 1974. Agricultural ecosystems. Agro-Ecosystems 1:1-6.

Higman, D., and D. L. Correll. 1982. Seasonal and yearly variation in meteorological parameters at the Chesapeake Bay Center for Environmental Studies. Pages 1-159 in D. L. Correll, editor. Environmental data summary for the Rhode River ecosystem, Volume A, Part I. Smithsonian Environmental Research Center, Edgewater, Maryland, USA.

Hutchinson, G. L., and A. R. Mosier. 1981. Improved soil cover method for field measurement of nitrous oxide fluxes. Soil Science Society of America Journal 45:311-316.

Jacobs, T. C., and J. W. Gilliam. 1985a. Headwater stream losses of nitrogen from two coastal plain watersheds. Journal of Environmental Quality 14:467-472.

Jacobs, T. C., and J. W. Gilliam. 1985b. Riparian losses of nitrate from agricultural drainage waters. Journal of Environmental Quality 14:472-478.

Kaplan, W. A., J. W. Elkins, C. E. Kolb, M. B. McElroy, S. C. Wofsy, and A. P. Duran. 1978. Nitrous oxide in fresh water systems: an estimate for the yield of atmospheric N_2O

associated with disposal of human waste. Pure and Applied Geophysics **116**:423–438.

Labroue, L., and G. Pinay. 1986. Natural nitrate removal from ground waters. Possibility of application to the reclamation of gravel pit lakes. Annales Limnologie **22**:83–88.

Lowrance, R., R. Todd, J. Fail, Jr., O. Hendrickson, Jr., R. Leonard, and L. Asmussen. 1984*a*. Riparian forests as nutrient filters in agricultural watersheds. BioScience **34**:374–377.

Lowrance, R. R., R. L. Todd, and L. E. Asmussen. 1984*b*. Nutrient cycling in an agricultural watershed: Part I. Phreatic movement. Journal of Environmental Quality **13**:22–27.

Lowrance, R. R., R. L. Todd, and L. E. Asmussen. 1984*c*. Nutrient cycling in an agricultural watershed: II. Stream flow and artificial drainage. Journal of Environmental Quality **13**:27–32.

Mooney, H. A., P. M. Vitousek, and P. A. Matson. 1987. Exchange of materials between terrestrial ecosystems and the atmosphere. Science **238**:926–932.

Naiman, R. J., H. Décamps, J. Pastor, and C. A. Johnston. 1988. The potential importance of boundaries to fluvial ecosystems. Journal of the North American Benthological Society **7**:289–306.

Peterjohn, W. T., and D. L. Correll. 1984. Nutrient dynamics in an agricultural watershed: observations on the role of a riparian forest. Ecology **65**:1466–1475.

Peterjohn, W. T., and D. L. Correll. 1986. The effect of riparian forest on the volume and chemical composition of baseflow in an agricultural watershed. Pages 244–262 *in* D. L. Correll, editor. Watershed research perspectives. Smithsonian Institution Press, Washington, D.C., USA.

Pinay, G., and L. Labroue. 1986. A natural sink for nitrates transported by ground water: the alder forest. Comptes Rendus des Seances De L'Academie des Sciences, Paris **302** (Series III):629–632.

Pinay, G., and H. Décamps. 1988. The role of riparian woods in regulating nitrogen fluxes between the alluvial aquifer and surface water: a conceptual model. Regulated Rivers: Research and Management **2**:507–516.

Risser, P. G., J. R. Karr, and R. T. T. Forman. 1984. Landscape ecology: directions and approaches. Illinois Natural History Survey, Special Publication Number 2. Champaign, Illinois, USA.

Wiens, J. A., C. S. Crawford, and J. R. Gosz. 1985. Boundary dynamics: a conceptual framework for studying landscape ecosystems. Oikos **45**:421–427.

RESTORATION OF HUMAN IMPACTED LAND-WATER ECOTONES

JAMES R. SEDELL, ROBERT J. STEEDMAN, HENRY A. REGIER AND STANLEY V. GREGORY. USDA Forest Service, Corvallis OR 97331 USA; Ontario Ministry of Natural Resources, Thunder Bay, Ontario, Canada P7C 5G6; The Institute for Environmental Studies, University of Toronto, Toronto, Ontario, Canada M5S 1A1; and The Department of Fisheries and Wildlife, Oregon State University, Corvallis OR 97331 USA.

Abstract. Scientific perspectives and management techniques applied to land-water ecotones in three regions of North America are reviewed. In the Pacific Northwest, the Arid West, and the Laurentian Great Lakes, the riparian and nearshore areas have functions that make them unusually valuable for fish and wildlife, commercial or aesthetic considerations. The science and practice of resource management, however, differs strongly in quality and quantity among these regions. Small-scale conflicts of resource allocation have begun to be addressed through local legislation and management policies, but attempts to coordinate effective basin-level and regional management approaches are as yet poorly developed. Much basic information about the ecological role of riparian and nearshore habitats is required, as are methods to study and model land-water interactions at medium to large scales of time and space. It seems likely that in spite of some useful generalizations that may apply at continental or global scales, much of the most important management information will be formulated and calibrated on a regional scale.

Key words: ecotones, Pacific Northwest, Arid West, Laurentian Great Lakes, restoration.

INTRODUCTION

In this essay we outline approaches and limitations to the management and rehabilitation of land-water ecotones in our regions of expertise. Recent experience in North America shows that there is strong need for practical understanding of the ecologic role of these boundaries, and of the way that management techniques must recognize regional differences in that role. In particular, we focus on hydrology, geomorphology and vegetation, as they determine the structure, dynamics and economic importance of the land-water boundary. We will argue for management perspectives that foster recognition of key features of land-water ecotones, so that the ecological processes centered there may persist in a sustainable manner. Policy makers need a list of options locating the best opportunities for protecting land-water ecotones, and for restoring degraded areas back to a self-sustaining state by incorporating desirable, natural and designed features.

Our essay briefly examines these issues in three regions of North America, to illustrate differences in scientific perspectives, natural resource policy and practical management approaches applied to protection and restoration of land-water ecotones. In the **Pacific Northwest,** values related to migratory salmonids in forested, high-gradient, wood-dominated river systems are predominant; in the **Arid West**, grazing and agriculture on intensely managed, water-limited, erosional watercourses, highlight the importance of riparian vegetation; in the **Laurentian Great Lakes**, commerce and urban development focused on nearshore and river-mouth features, have resulted in basin-level rehabiliation initiatives to restore lost biotic, cultural and economic values of shorelines and tributaries.

Key ecological features of a land-water interface are not distributed evenly in space along the ecotone. Rather they are often localized areas of hydrologic stability and sediment deposition, as may occur at channel confluences, wide floodplains or river mouths. These areas change structurally through time, but show common characteristics regardless of whether they are formed along the edges of rivers or lakes. The most important of these characteristics are extensive surface and subsurface hydrologic connectivity of the ecotone with the adjacent upland and aquatic systems; diverse vegetation represented by a variety of age classes; and resilience in the response to changes in the hydrologic and geomorphic characteristics of the boundary.

The science and management of riparian resources seems most developed in the Pacific Northwest, and this is where we are best able to provide practical detail. The other regional examples illustrate the exciting conceptual parallels and management innovations that are being applied in widely divergent landscapes of North America.

RIVERINE RIPARIAN ZONES IN FORESTED AREAS OF THE PACIFIC NORTHWEST

Regional Features

From an ecological perspective, riparian zones along rivers can be defined functionally as three-dimensional zones of direct interaction between terrestrial and aquatic ecosystems (Figure 1). The boundaries of riparian zones, as defined from this perspective, extend outward from the channel to the limits of flooding and upward into the canopy of streamside vegetation (Meehan et al. 1977, Swanson et al. 1982). Dimensions of a zone of riparian influence vary in relation to the dynamics of particular processes of interest. For example, in a coniferous forest, the zone from which large woody debris could enter the active channel or floodplain may extend farther away from the stream than the zone from which leaves and needles would be contributed. The zone of shading influence may be asymmetrical, depending on stream orientation, latitude, and topographic shading. In the Pacific Northwest, the riparian three-dimensional zone occupies an area of less than 5 percent of the total landscape.

Development of an ecosystem perspective of riparian zones must incorporate the complex array of physical, chemical, and biological interactions that occur within the interface between aquatic and terrestrial ecosystems. Of key importance here is the geomorphic context that valley floor landforms provide for riparian zones. Stream reaches are delineated by the type and degree of physical constraint imposed by the valley wall. The degree of constraint controls local geomorphic processes, and therefore influences terrestrial and aquatic communities through erosion, sedimentation, soil movement and other disturbance mechanisms.

Constrained reaches in which the valley floor is narrower than two active stream channel widths are formed where bedrock, landslides, alluvial fans, or other geologic or human-produced features constrict the valley floor, and thus limit lateral mobility of the channel. Streams within constrained reaches tend to be relatively straight, single channels with limited lateral heterogeneity. During high flow, the position of the constrained stream channel is relatively fixed within narrow floodplains and stream power increases rapidly with increasing discharge. Relative resistance to erosion affects the persistence of the constraint and the composition of the substrata in the active channel. Valley floors in constrained reaches are characteristically narrow and include few geomorphic surfaces within the valley floor. Riparian vegetation in these areas is usually similar in composition to adjacent hillslope plant communities.

Unconstrained reaches occur in areas in which the valley floor is wider than two active channel widths affording some lateral mobility to the channel. Streams within unconstrained reaches tend to be complex, multiple channels with diverse ecotonal heterogeneity. During floods, the positions of the unconstrained stream channel is dynamic within the floodplain and stream power can dissipate by flooding over numerous geomorphic surfaces within the valley floor with increasing discharge. There is stream bank and bed erosion in these dynamic areas. Riparian vegetation in these

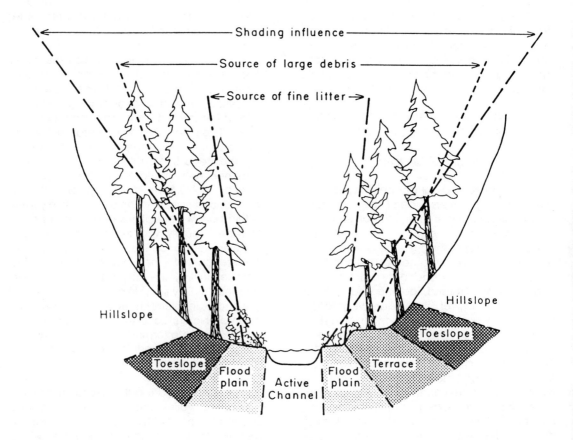

Figure 1. Cross-sectional view of valley landforms and associated riparian plant communities, illustrating varying zones of terrestrial influence on aquatic ecological processes (modified from Gregory et al. *in press*).

areas is diverse both in terms of species composition and age classes within a species.

Unconstrained reaches in the Northwest region of North America are ecological "hotspots." They are areas which retain organic materials, have much greater algal primary production, have the highest densities and diversity of fishes, greater aquatic invertebrate densities, and much greater microbial activity as indicated by ammonium uptake (Gregory et al. *in press*). These are areas of greatest subsurface flow and connectivity with surface water and greatest interaction with the adjacent forest. Generally, in the Western Cascades of Oregon the percentage of channel length in an unconstrained condition does not exceed 25% in 4-5th order watersheds. However, the stream banks of unconstrained reaches have 70-150 percent more edge than banks of constrained reaches. Thus, the ecological value of unconstrained reaches depends primarily on physical configuration and land-water linkages, and is generally independent of spatial scale. In the Pacific Northwest, resource managers with limited money need to focus their best management practices and restoration efforts preferentially in these areas.

Designed Ecotone Structure and Edge Complexity

Streams in forested ecotones need a continuous supply of large woody debris to maintain complexity of aquatic habitats. In-stream channel structures or debris substitutes that are used as surrogates for large trees in ecotones and stream channels are conceptually popular with both timber and fisheries managers. The fisheries enhancement programs of many agencies emphasize construction of boulder berms, gabion structures of various configurations, boulder clusters, side channels, and off-channel ponds as primary techniques for habitat improvement on stream channels. Stream habitat enhancement projects on the west coast have improved spawning and rearing habitat for salmonids and probably increased fish production in some areas. However, rigorous ecological evaluation of such projects is rarely undertaken to verify benefits.

Economic analysis of enhancement projects shows that substituting other structures for large woody debris can be expensive. The cost of placing individual boulders in the Keogh River, of British Columbia for example, was $22 to $24 per boulder (in 1977 Canadian dollars), depending on whether the boulders were placed with heavy equipment or by helicopter (Ward and Slaney 1979). Planning estimates of costs for boulder placement in western Oregon streams average $35 per boulder (House and Boehne 1985). Installing gabions for spawning gravel retention in tributaries of the Coos River, Oregon, cost a total of $225,000 in 1981. Cost per individual gabion ranged from $300 in a fourth-order stream to $1,700 in a sixth-order stream (Anderson 1982). An average cost of $1,200 for material and labour per gabion does not include engineering design or road access costs (House and Boehne 1985). To adequately restructure one mile of stream costs between $12,000 and $20,000, with a possible project longevity of 20 to 30 years.

Forest managers must weigh the cost of enhancement structures against the cost of providing woody debris to streams through forest management in riparian zones. In many instances, after-the-fact substitution of debris may cost much more than allowing debris to be recruited to the channel naturally (House and Crispin 1990). Where structural enhancement is warranted, use of native materials such as logs, may be the most cost-effective means of achieving the hydraulic diversity necessary for productive fish habitat (Lisle 1982).

Another problem with fisheries programs designed around enhancement structures is the general difficulty in finding suitable construction sites. From a logistic standpoint, not all stream miles could be improved with enhancement devices even if unlimited funding was available. For example, one

Bureau of Land Management District in Oregon has 528 miles of streams producing anadromous fish, and only 60 miles are feasible for structural rehabilitation. Approximately 90 percent of the stream miles cannot be "artificially" rehabilitated because of access problems and other constraints, and must be managed for vegetative diversity in forested ecotones to maintain acceptable levels of fish production.

Enhancement structures are going to be placed in streams at an accelerated rate into the 1990's. Although structural additions and alterations may improve habitat in streams lacking physical diversity, they cannot replicate the dynamic physical interactions between water, sediment and riparian vegetation. Structures are usually a short-term solution, and are implemented in the absence of a long-term watershed improvement plan.

In forested basins, comprehensive watershed planning must call for reestablishment of woody vegetation and accelerated growth of trees, to provide snags and wood to the stream. Such silvicultural management programs must be key components of fish and wildlife habitat improvement programs, but silvicultural prescriptions designed to grow streamside vegetation are neither widely used nor well understood. Natural riparian forest can foster structural diversity by providing dead wood, varying canopy heights, and a varied vegetative community. However, maintaining biological diversity and habitat for fish and wildlife species involves much more than stream-side strips of riparian vegetation. Research can tell us how various tree species, root systems and debris sizes function through time in channels of different widths, gradients, and valley form. Such information is necessary to aid decisions regarding where, what kind, and how many trees are to be selectively planted or maintained in riparian areas to optimize ecotone structure and function over the long term. Management policies and practices are needed that will allow riparian vegetation to maximize its beneficial effects

upon the hydrology and channel morphology of stream and lake ecosystems.

Basin Planning for Maintenance of Ecological Hotspots

The movement of water and sediment downwards through a drainage network translates detrimental effects of forestry off-site and into aquatic environments. In the Pacific Northwest, with its steep terrain and high precipitation, spatial and temporal patterns of runoff and soil movement are important in determining the extent and severity of aquatic habitat degradation. New techniques allow areas with high rates of natural sediment production from debris slides or earth flows to be identified within a drainage basin, so that careful basin planning can minimize damaging effects of forestry operations on fishery resources (Megahan and King 1985, Sidle et al. 1985, Swanston 1985). In areas with high relief and abundant stream channels, knowledge of the movement of debris flows through drainage networks allows managers to maintain the structure and function of ecological hotspots along stream edges (Benda 1985, Swanson et al. 1987).

Standards and Guidelines for Riparian Management Areas in the Pacific Northwest

The important functional role of riparian forests in providing trees to streams and rivers has been recognized by land managers in the western United States (Bisson et al. 1987, Sedell et al. 1988). All western state and federal jurisdictions have adopted standards for leaving large trees in riparian zones to protect fish and wildlife. However, management schemes to maintain diverse ecotones in the Pacific Northwest have been difficult to develop because of conflicting resource demands in riparian areas and the shortage of information on silvicultural management strategies for riparian zones.

Most streamside management schemes incorporate an un-harvested buffer strip of trees along the stream channel. The width of the buffer, and the age and abundance of its trees are important ecologic and economic concerns. Reduced harvest frequency in riparian zones may protect important riparian functions, by increasing average tree age. Large, old wood is required in stream channels to provide structural complexity and aid in formation of pools and complex channel edges. Further, the diameter and length of wood must increase with increasing stream size to be effective in formation of good fish habitat. Most National Forests in western United States have planned some form of double rotation scheme for stream ecotones. The rationale for a double rotation comes from the fact that large wood >30 cm in diameter is required in stream channels to provide structural complexity and aid in formation of pools and edge complexity. Woody debris at least 60 cm in diameter and 13 m in length is required on coastal streams with widths greater than 10 m. Wood recruitment studies by Grette (1985), Long (1987) and Heinman (1988), indicate that large coniferous debris does not enter streams in significant amounts until the stand age reaches 120 to 150 years.

Studies that integrate riparian silvicultural strategies for fish habitat considerations are rare. A study by Rainville et al. (1985), however, attempted this kind of integration for the Idaho Panhandle National Forest. The investigators linked stream habitat inventory information and silvicultural data and developed a computer simulation model to evaluate the effect of silvicultural practices on the potential recruitment of large wood pieces to stream channels. Forest habitats of western hemlock (*Tsuga heterophylla*), grand fir (*Abies grandis*), and subalpine fir (*Abies lasiocarpa*) were analyzed. Active management was clearly essential to maximize stream habitat complexity and fisheries resources in these vegetative habitat types.

The "postage stamp" entry technique of Rainville et al. (1985) where 200 m long blocks are clearcut in ecotones is a marked contrast from the forest practice rules of Oregon and Washington, and the general guidelines used in the National Forests of the Pacific Northwest and Alaska, which generally call for a selective harvest of large trees in forested ecotones. The selective harvest approach would ideally leave a stream with a canopy of diverse vegetative structure and species along its entire length. The postage stamp approach optimizes fish habitat structure over the entire basin, and requires a detailed basin harvest and silvicultural plan.

The cost of leaving economically valuable timber in the ecotone varies as a function of tree abundance, but is usually relatively less expensive than equivalent upland reserves. In general, the densities of large conifers in forested ecotones are 70% of upland densities and basal area in the western Cascades (Tom Spies, *personal communication*) and 15 to 50% of the densities and basal area in the Oregon Coast range (Oregon State Department of Forestry 1987, Andrus and Froehlich 1988). If all of the 3rd order fish-bearing streams in the upper Willamette River basin were protected by not harvesting large conifers within 30 m of the stream edge, the resulting loss of available timber in the basin would amount to only 0.3 percent.

House and Crispin (1990) did an economic analysis of the value of large wood debris from the ecotone forest to salmonid habitat in coastal Oregon streams. They showed that the fisheries benefits of maintaining conifers in the ecotone at a high rate of wood loading levels were calculated to 11 percent greater by year 20, and 59 percent higher after 94 years (an increase of over $50,000 in present value per kilometer of stream) over the stump value of the conifers in the ecotone. Management scenarios for structural stream rehabilitation versus managing for conifer densities in the ecotone showed greater short term values for structural rehabilitation efforts. However, long term economic benefits of rehabilitation management were substantially less than for those streams managed under continuous high debris loading.

115

RIPARIAN ECOTONES IN ARID AND AGRICULTURAL LAND

Many groundwater aquifers associated with river channels in the western United States are maintained by infiltration of upland runoff. These alluvial aquifers are an important source of water for human use and for riparian vegetation. Water storage in such aquifers was once partially responsible for maintaining baseflow in western rivers, many of which are now dry during much of the year. Removal of riparian vegetation has been responsible in part for the change from perennial to intermittent flow in some of these rivers (McNatt 1978, Elmore and Beschta 1987).

Riparian Values and Management

Figure 2A provides a cross-sectional view of a small, degraded stream system in an arid landscape. In this example, the stream has cut down through previously deposited alluvium. As a result, the channel and associated vegetation have changed dramatically. Plant species typical of wetland conditions have largely disappeared and the channel continues to erode laterally. There is little subsurface storage of water and the stream is characterized by intermittent flow.

In contrast, Figure 2B illustrates a previously eroded channel that supports a diversity of riparian vegetation and has undergone recovery. The vegetation provides relative stability to stream banks and causes deposition of sediment. Over time the channel undergoes aggradation. Such aggradation is often a natural consequence of allowing streamside vegetation modified by grazing, logging, agriculture or other management practices, an opportunity to recover. A consequence of this aggradation process is that the water table will similarly rise. In some cases, a formerly intermittent stream may flow perennially.

Agriculture and management of riparian vegetation for erosion control can be compatible along floodplain systems. For centuries, native American and Spanish American farmers of the arid Southwest have managed riparian vegetation adjacent to their agricultural fields (Nabhan 1985). They planted, pruned, and encouraged tree species for flood erosion control, soil fertility renewal, buffered field microclimates, and fuel wood production. Living fence rows were constructed by weaving brush between the trunks of lines of cottonwood, willow, and mesquite adjacent to their floodplain field. This woven fence slowed lateral flood waters without channelizing the primary streambed the way in which concrete or riprap channel banks would. As a result, when summer or winter floods covered a floodplain terrace, channels are less likely to become entrenched and erosion is less pervasive than with a barren floodplain (Nabhan 1985).

Streamside vegetation also maintains and enhances water quality in streams draining agricultural lands (see Correll this volume). In the midwest and southeastern coastal plain, woody riparian vegetation not only stabilizes banks and creates complex fish habitat, but it also filters nutrients and maintains water quality on agricultural watersheds. The removal of nutrients such as nitrogen and phosphorus occurs via several mechanisms (Lowrance et al. 1985, Petterjohn and Correll 1984): (1) surface filtration of sediments, (2) incorporation of N and P into living woody plants; and (3) nitrification-denitrification processing below ground and at the soil surface. Soils of the riparian ecosystem present ideal conditions for denitrification: high organic matter from input of forest litter; seasonal waterlogging; and large inputs of nitrates in subsurface flow. Most of the nitrogen goes to the atmosphere as gas via denitrification with only a small amount incorporated into the biomass of the growing trees. In Maryland, a 15 m buffer-strip of trees is required between agricultural land and the Chesapeake Bay and adjacent tributaries. This width of a filter strip can remove more than 75 percent of the ground water nitrogen and more than 40 percent of the phosphorus before it gets into the adjacent stream or water body.

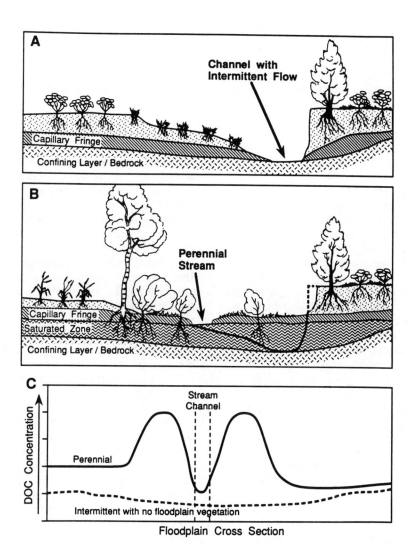

Figure 2. General characteristics and functions of riparian areas in arid lands (from Elmore and Beschta 1987). (A) Degraded riparian area with very little surface and subsurface water connectivity and no vegetation interaction with the stream. (B) Recovered riparian area with increased connectivity of subsurface and surface water, vegetation stabilized banks and extensive interaction with diverse vegetation. (C) Dissolved organic carbon (DOC) concentrations in floodplain groundwater, along transects for perennial and intermittent streams.

Currently trees in ecotones of arid lands play a limited role in agriculture. Benefits derived from a dense, deep-rooted tree buffer strip may include:

1. Production of a woody crop entirely useful as a renewable fuel or pulp wood.

2. Roots and root exudates will place more organic carbon deeper in the riparian zone soil profile (Figure 2C). Theoretically, an increased amount of organic carbon affects pesticide concentrations in groundwater by increasing the rate of pesticide degradation by microbes within the soil profile.

3. Biological diversity enhanced by habitat restoration. Fish and wildlife habitat is usually highly fragmented in agricultural watersheds. Restoration of corridors linking alternative habitats will enhance the health of both aquatic and terrestrial organisms.

The economic value of forested riparian corridors in floodplains of the agricultural midwest has been recently addressed by Lant and Tobin (1989). They illustrate how an economically efficient mix of woody riparian corridors and cropland can be estimated and they suggest how such a mix of land uses can be encouraged through appropriate agricultural policies (i.e., economical incentives). They also emphasize that the location of riparian wetlands and corridors within an agricultural watershed greatly influences their effectiveness in providing the environmental services discussed above.

The placement of riparian forest corridors in agricultural land will often be dictated by planning for flood control, sediment control, groundwater management and recreational benefits. In the longitudinal dimension, riparian forests cannot directly impact flooding and water quality upstream of their occurrence, and their influence will gradually diminish in a downstream direction. Thus, riparian forests and wetlands in lower reaches of the fluvial system, or directly upstream of valued recreation areas or reservoirs, may provide more benefits than similar sized riparian forest or wetlands elsewhere in the watershed (Johnston et al. 1990). In the lateral dimension riparian forests adjacent to the riverbank or in areas of high probability of flooding will have the greatest beneficial impact on water quality and flooding.

SHORELINES IN GREAT LAKES ECOSYSTEMS

The shorelines and watersheds of the Laurentian Great Lakes have been divided between the Province of Ontario and the States of Minnesota, Wisconsin, Illinois, Indiana, Ohio, Pennsylvania and New York. More than 36 million people live in the Great Lakes basin, and are concentrated in a southern megalopolis comprising the cities of Chicago, Detroit, Toledo, Cleveland, Buffalo, Hamilton, Toronto, Rochester, and others. The southern shores and tributaries, along Lakes Michigan, Erie and Ontario, have been widely restructured and degraded in the face of extensive urbanization and industrialization. In the north, Lakes Huron and Superior are sparsely populated, with development and industrialization concentrated along a few large bays and river mouths.

The ecological characteristics of riparian and nearshore ecosystems in the Great Lakes are analogous in many ways to marine systems such as the Baltic Sea for large-scale features (Regier et al. 1988), and to inland lakes and rivers, for small-scale features (Steedman and Regier 1987). Governance and management of Great Lakes land-water ecotones is complicated by the large number of federal, state/provincial, regional and municipal agencies with mandates relevant to riparian land, navigable waters, conservation, environmental protection, fisheries and wildlife. The binational and regional approaches that have been developed for the protection, rehabilitation and management of Great Lakes land-water ecotones are among the most ambitious and complex that have been attempted for such a resource.

Regional Ecosystem Features

Key habitats.–Certain locales or habitats along the land–water ecotone may be more important than others to the health and survival of Great Lakes aquatic ecosystems (Steedman and Regier 1987). Relatively small or localized areas that provide essential conditions for breeding, spawning, rearing, and feeding of fishes may have an ecological role far more important than would be suggested by their size alone. In temperate aquatic ecosystems such as the Laurentian Great Lakes, these "centers of organization" tend to occur in the coastal or nearshore areas, and exhibit distinctive combinations of abiotic and biotic characteristics.

In the Great Lakes Basin, the structural and hydrologic features of river channels, coastlines, rocky shoals, estuaries and coastal wetlands are such that in a natural condition: 1) substrate and sediment accumulations are of a size and arrangement that either provides clean, well oxygenated substrate surfaces and interstices, and/or sediments suitable for the establishment of aquatic plants; and 2) disruptions by currents, wave action, ice movement, seiches or floods are of a frequency, intensity and predictability that allows a variety of plants and animals to colonize the area, either for vulnerable embryonic or juvenile stages, or for the entire life cycle. With respect to native, valued species of Great Lakes fishes, the most important requirements of the early life stages, i.e. provision of oxygenated water and protection from predation, are best met in these habitats.

Quantity and Quality of Great Lakes Land–Water Ecotones.–In urbanized areas of the Great Lakes there has been widespread destruction of nearshore and tributary structural features through activities such as shoreworks, channelization, land clearing, landfilling, land drainage, rock removal, dredging, and siltation. The remaining locales that seem to function as ecological

"centers of organization" are indeed becoming identifiable as discrete and often isolated entities, probably much more so than would have been the case 200 years ago when such features dominated much of the Great Lakes nearshore ecotone.

The destruction of coastal wetlands, once widespread on the lower Great Lakes, serves to illustrate this point. Most authorities estimate that about two–thirds of those wetlands have been filled or drained for agricultural, industrial and urban development (Patterson and Whillans 1985, Conservation Foundation and Institute for Research on Public Policy 1989). Whillans (1984) summarized physical characteristics of the Canadian shoreline of the Great Lakes (Table 1) and found that only about 30% overall of the Canadian shoreline provides sheltered embayments or estuaries which may support wetland development and growth of aquatic plants. The rest of the shoreline is relatively exposed and wave-swept, but may have important features such as gravel beaches or bedrock shoals required for spawning of salmonids and coregonids. Shoals large enough to be mapped on navigational charts are surprisingly rare even on the rugged shorelines of Lake Superior and Lake Huron.

Information about the quality, distribution and ecological importance of wetlands and other key shoreline habitats is being gathered by provincial and state agencies to facilitate protection and rehabilitation. This information has not yet been compiled and mapped on a basin scale, although steps have been made in that direction (Patterson and Whillans 1985, Botts and Krushelnicki 1988).

With regard to potential use by valued fish species, some broad–scale indication of the ecological nature of Great Lakes tributaries may be available through the Great Lakes Fisheries Commission, as part of their sea lamprey control program. The sea lamprey (*Petromyzon marinus*) feeds on a

Table 1. Physical characteristics of Great Lakes Shoreline (Canadian side only) (Whillans 1984)

SHORELINE LENGTH (km)

Shoreline Type	Ontario	Erie	St. Clair	Huron	Superior	Total
Exposed	472	493	47	1254	936	3202
Embayment	68	107	20	396	106	697
Estuary	29	1	5	265	49	349
Protected (Manmade)	35	47	61	49	36	228
Total	604	648	133	1964	1127	4476
No. of Shoals	34	19	25	339	67	484

variety of valued fish species, and is the object of a multi-million dollar control effort on the Great Lakes. The sea lamprey spawns in certain streams and rivers tributary to the Great Lakes, and in some nearshore areas off large river mouths. Streams that support larval lamprey are generally suitable for spawning by "anadromous" salmonids, by virtue of migratory access, cool temperatures and good water quality. Of approximately 5747 streams tributary to the Great Lakes (Table 2), only 433, or 7.5%, are known to have supported sea lamprey since 1957 (Mormon et al. 1980). Of these, 364 are on the upper Lakes (Superior, Huron, Michigan), and 69 are on Lakes Erie and Ontario. In spite of the large number of tributary streams on the Great Lakes, waters suitable for reproduction and rearing of valued fish species may be relatively rare. If one assumes that the average width of a river mouth on the Great Lakes is 0.1 km, then approximately 570 km or 3.4% of the Great Lakes shoreline is "riverine." If one further assumes that use of a river by sea lamprey denotes the presence of salmonid habitat, then only 43 km or 0.25% of the Great Lakes shoreline provides access to high quality habitat for riverine salmonids.

Rehabilitation techniques for Great Lakes Land-Water Ecotones

Techniques and feasibility of ecosystem rehabilitation in nearshore waters of the Great Lakes were examined in detail at a 1979 workshop in Toronto, Ontario, convened under the auspices of the Great Lakes Fisheries Commission. Here, rehabilitation connotes a pragmatic mix of restoration, enhancement, remediation, mitigation and preservation, to benefit highly valued, sensitive uses of Great Lakes ecosystems. Approximately 50 recognized experts in the natural science and engineering of ecological rehabilitation pooled their ideas to produce the synopsis documented in Francis et al. (1979). That synopsis, while too lengthy to include in this volume, represents one of the most comprehensive and detailed listings of practical rehabilitation techniques compiled for the Great Lakes, and should be examined by readers interested in further detail.

Francis et al. (1979) listed 18 human-caused ecological stresses in Great Lakes ecosystems; of these 18 stresses, 11 are clearly relevant to a consideration of nearshore or coastal degradation, and are listed in Table 3. Although the synopsis of Francis et al. (1979) is now more than ten years old, it appears to have stood the test of time, particularly with regard to major institutional rehabilitative efforts. A contemporary workshop to address these issues would probably provide a number of innovative ideas and approaches to ecosystem rehabilitation, most likely with regard to handling and treatment of toxic substances. However, most of the approaches to structural rehabilitation and remediation of conventional pollution would be substantively similar.

It was not possible to estimate rehabilitation costs in 1979; this is likely still the case. Nonetheless, Francis et al. (1979) reached two conclusions which are still valid. First, a great deal is not known about the benefits and costs of rehabilitation measures. Second, it is readily apparent that some degree of rehabilitation will be highly beneficial to all users of the Great Lakes and their ecotones.

Remedial Action Plans on the Great Lakes

The Great Lakes Water Quality Agreement of 1978, amended in 1987 (International Joint Commission 1988), provides a key binational vehicle for rehabilitation and restoration of land-water ecotones on the Great Lakes. Under the Agreement, the Federal governments of Canada and the United States, in cooperation with provincial and state agencies, committed to develop and implement Remedial Action Plans (RAPs) for 42 Areas of Concern on the Great Lakes and their connecting channels (Figure 3). Areas of Concern were defined as locations, usually heavily industrialized river mouths or embayments, where environmental

Table 2. Great Lakes shoreline lengths and tributary counts.

Lake	Shoreline Length (km)[1]	Number of Tributaries[2,3,4]**	Stream km below first barrier (Canadian only)[5]
Superior	4385	840	1091
Huron	6157	1654	1720
Michigan	2633	447	
Erie	1402	558	259
Ontario	1146	507	246
Canadian Shoreline U.S. shoreline		2860 1662	33164
Total	17017*	5747	

* includes connecting channels
** different subtotals from different authorities

[1]Botts and Krushelnicki (1988)

[2]Lawrie and Rahrer (1973)

[3]Mormon et al. (1980)

[4]Christie (*personal communication*)

[5]Bird and Rapport (1986)

Table 3. Ecological stresses active in nearshore or coastal areas of the Laurentian Great Lakes (Francis et al. 1979).

---

1. Microcontaminants: toxic wastes and biocides
2. Nutrients and eutrophication
3. Organic inputs and oxygen demand
4. Sediment loading and turbidity
5. Stream modification: dams, channelization, logging and changes in land use
6. Dredging and mineral, sand, gravel and oil extraction
7. Filling, shoreline structures, offshore structures
8. Water level fluctuation and control
9. Dyking and hydraulic modifications of wetlands
10. Thermal loading
11. Major degradative incidents or spills

Great Lakes Areas of Concern

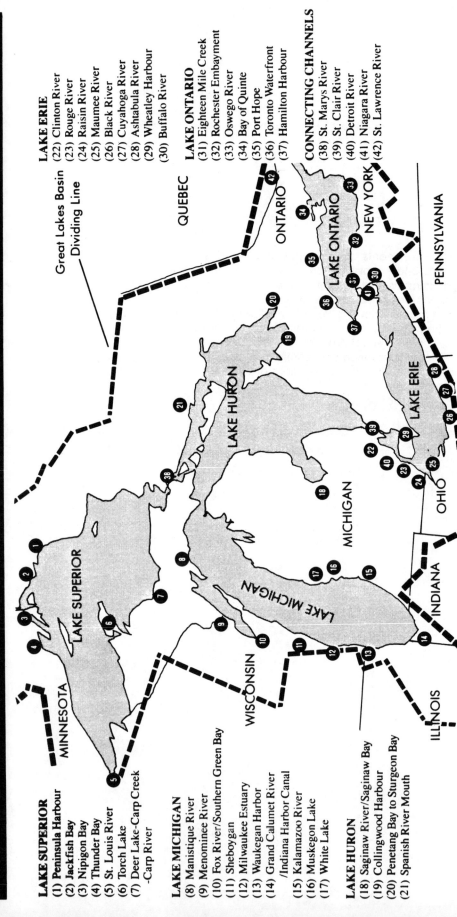

LAKE SUPERIOR
(1) Peninsula Harbour
(2) Jackfish Bay
(3) Nipigon Bay
(4) Thunder Bay
(5) St. Louis River
(6) Torch Lake
(7) Deer Lake-Carp Creek
    -Carp River

LAKE MICHIGAN
(8) Manistique River
(9) Menominee River
(10) Fox River/Southern Green Bay
(11) Sheboygan
(12) Milwaukee Estuary
(13) Waukegan Harbor
(14) Grand Calumet River
    /Indiana Harbor Canal
(15) Kalamazoo River
(16) Muskegon Lake
(17) White Lake

LAKE HURON
(18) Saginaw River/Saginaw Bay
(19) Collingwood Harbour
(20) Penetang Bay to Sturgeon Bay
(21) Spanish River Mouth

LAKE ERIE
(22) Clinton River
(23) Rouge River
(24) Raisin River
(25) Maumee River
(26) Black River
(27) Cuyahoga River
(28) Ashtabula River
(29) Wheatley Harbour
(30) Buffalo River

LAKE ONTARIO
(31) Eighteen Mile Creek
(32) Rochester Embayment
(33) Oswego River
(34) Bay of Quinte
(35) Port Hope
(36) Toronto Waterfront
(37) Hamilton Harbour

CONNECTING CHANNELS
(38) St. Marys River
(39) St. Clair River
(40) Detroit River
(41) Niagara River
(42) St. Lawrence River

Great Lakes Basin
Dividing Line

Figure 3. International Joint Commission "areas of concern" in Laurentian Great Lakes coastal ecotones (map courtesy of The Great Lakes Reporter, The Center for the Great Lakes, Chicago, Illinois).

objectives of the Agreement were not being met and where aquatic life was impaired.

The language and intent of the Great Lakes Water Quality Agreement are strongly consistent with current ecological perspectives on land-water linkages, and emphasize the great importance of the land-water ecotone. Annex 2 of the Agreement states that Remedial Action Plans are to serve as an important step toward virtual elimination of persistent toxic substances, and toward restoring and maintaining the chemical, physical and biological integrity of the Great Lakes Basin Ecosystem.

The spatial context of Remedial Action Plans is broadened by the ecosystemic perspective of the Great Lakes Water Quality Agreement. In practice, many RAPs are focusing not only on nearshore and shoreline habitats, but are looking upstream into the watersheds for solutions to poor land-use practices, urban stormwater and pollutants with non-point source origins. In the process of gathering and compiling recent and historical information about environmental problems in the areas of concern, a much improved picture of local and regional land-water interactions is emerging.

The 42 Areas of Concern, and many others not officially recognized, are or have been important centers of ecological organization, not least in that they have provided spawning, rearing and feeding habitat for a wide variety of desirable fish species. The characteristics that made these areas biologically important (sheltered waters and access to river mouths in particular) also made them centers of cultural organization, as harbours and centers of trade. With rapid growth of urban re-development, sport fisheries and tourism in the Great Lakes basin in recent years, it has become apparent that there is both an economic and an ecologic imperative to restore the integrity and attractiveness of these unique ecotones.

Perhaps one of the most important aspects of Remedial Action Planning in the Great Lakes basin is the requirement for comprehensive public consultation (i.e. Eder and Jackson 1988) in both the identification of impaired uses in the areas of concern, and in the restoration of impaired uses. Strong governmental commitment to success of RAPs can only be made stronger with the development of an informed, interested constituency for cleanup and rehabilitation of the Great Lakes and their shoreline ecotones.

CONCLUSIONS

Management strategies for streamside and lakeside ecotones on public and private lands have begun to shift in recent years and continued change is expected. Because of the importance of these areas for a wide variety of values (i.e., water quality and quantity, aesthetics, fisheries, wildlife, forage, timber, and recreation), the general public is increasingly involved in riparian and wetland regulation and policy decisions. Thus, lakes, streams, and their adjacent lands have become a focal point of intense interest, and sometimes of conflict, for a wide variety of groups.

Perspectives of land-water ecotones based on isolated components of the terrestrial-aquatic interface are ecologically incomplete, and have limited application to practical understanding and intelligent, sustainable management of natural and designed ecosystems. The complex mosaics of landforms and terrestrial and aquatic communities within river valleys and along lakeshores are important, dynamic components of the landscape. Management of these resources requires a conceptual framework that integrates physical and biotic processes responsible for valley floor landforms, river channels, coastal zones and patterns of terrestrial plant succession.

The examples of land-water ecotone management presented here were chosen to illustrate the integrated, self-organizing components of this valuable landscape feature. Increased understanding of landscape and ecological process has increased the number of management options currently available for the conservation and restoration of land-water ecotones. Further understanding of locally and regionally important ecological processes at these ecosystem boundaries, and identification of ecologically important features along river systems and coastlines will foster development of additional management options.

Defining economic values associated with improved environmental services and non-consumptive uses of riparian areas represents a challenge that may extend well beyond the present benefit-cost paradigm of natural resource economists. The development of informed environmental constituencies must precede the coordinated movement of hundreds of local, state or provincial, national, and international regulatory jurisdictions toward common environmental goals. Such progress in our way of doing things clearly depends on formulation of ecologically relevant natural resource and economic policies. This must be guided by ecosystem science that fosters regional understanding of ecosystem function. Success in this endeavour at the land-water ecotone may well constitute success in the broadest sense. Practical expertise of broad ecological relevance will be developed through successful ecotone management. Of perhaps more importance in many regions, land-water ecotones represent highly sensitive landscape units. Sustainable protection there will protect key processes and features on adjacent, less sensitive upland and aquatic habitats.

ACKNOWLEDGEMENTS

The authors thank M.M. Holland, M.N. Jensen, R.J. Naiman, P.G. Risser, and an anonymous reviewer for comments on earlier drafts of the manuscript. This work was supported in part by National Science Foundation (NSF) grants BSR-8414325 and RSR-8508356 to Oregon State University (for research at the H.J. Andrews Experimental Forest), the Ontario Ministry of Natural Resources ORRRGP GR-050-84, and the Ontario Ministry of the Environment.

LITERATURE CITED

Anderson, J. W. 1982. Anadromous fish projects of 1981, U.S. Department of the Interior - Bureau of Land Management, Coos Bay District. Pages 109-114 in T. J. Hassler, editor. Proceedings on the propagation, enhancement, and rehabilitation of anadromous salmonid populations and habitat in the Pacific Northwest. California Cooperative Fisheries Research Unit, Humboldt State University, Arcata, California, USA.

Andrus, C., and H. Froelich. 1988. Streamside forest development following logging or fire in the Oregon Coast Range: wildlife habitat provided and value of included timber. Pages 139-152 in K. J. Raedeke, editor. Proceedings of the Riparian Wildlife and Forestry Interactions Symposium. University of Washington, Seattle, Washington, USA.

Benda, L.E. 1985. Delineation of channels susceptible to debris flows and debris floods. Pages 195-201 in International symposium on erosion, debris flow, and disaster prevention. Tsukuba, Japan.

Bird, P.M. and G.J. Rapport. 1986. State of the environment report for Canada. Minister of Supply and Services Canada, Catalogue Number EN 21-54/1986E, Ottawa, Canada.

Bisson, P.A., R.E. Bilby, M.D. Bryant, C.A. Dolloff, G.B. Grette, R.A. House, M.L. Murphy, K.V. Koski, and J.R. Sedell. 1987. Large woody debris in forested streams: past, present and future. Pages 143-190 in E.O. Salo and T.W. Cundy, editors. Streamside management: forestry and fishery interactions. University of Washington Institute of Forest Resources Contribution Number 57. University of Washington, Seattle, Washington, USA.

Botts, L., and B. Krushelnicki. 1988. The Great Lakes: an environmental atlas and resource book. Environment Canada, Great Lakes Environment Program, Catalogue Number EN40-349/1987E, Toronto, Ontario, Canada and United States Environmental Protection Agency, Great Lakes National Program Office, EPA-905/9-87-002 GLNPO NO-2, Chicago, Illinois, USA.

Conservation Foundation and Institute for Research on Public Policy. 1989. Great Lakes: great legacy? Conservation Foundation and Institute for Research on Public Policy, Baltimore, Maryland, USA.

Correll, D.L. 1991. Human impact on the functioning of landscape boundaries, this volume.

Eder, T., and J. Jackson. 1988. A Citizen's Guide to the Great Lakes Water Quality Agreement. Great Lakes United, State University of New York, Buffalo, New York, USA.

Elmore, W., and R.L. Beschta. 1987. Riparian areas: perceptions in management. Rangelands 9(6):260-265.

Francis, G.R., J.J. Magnuson, H.A. Regier, and D.R. Talhelm. 1979. Rehabilitating Great Lakes ecosystems. Great Lakes Fishery Commission Technical Report 37:1-99.

Gregory, S.V., F.J. Swanson, and W.A. McKee. In press. An ecosystem perspective of riparian zones. Bioscience.

Grette, G.B. 1985. The abundance and role of large organic debris in juvenile salmonid habitat in streams in second growth and unlogged forests. Masters Thesis, University of Washington, Seattle, Washington, USA.

Heimann, D.C. 1988. Recruitment trends and physical characteristics of coarse woody debris in Oregon coast range streams. Masters Thesis, Oregon State University, Corvallis, Oregon, USA.

House, R.A., and P.L. Boehne. 1985. Evaluation of instream enhancement structures for salmonid spawning and rearing in a coastal Oregon stream. North American Journal of Fisheries Management 5:283-295.

House, R., and V. Crispin. 1990. Economic analyses of the value of large woody debris as salmonid habitat in coastal Oregon streams. United States Department of the Interior Bureau of Land Management Technical Note T/N OR-7, Portland, Oregon, USA.

International Joint Commission. 1988. Revised Great Lakes Water Quality Agreement of 1978, as amended by Protocol signed November 18, 1987. Consolidated by the International Joint Commission, United States and Canada, Windsor, Ontario.

Johnston, C.A., N.E. Detenbeck, and G.J. Niemi. 1990. The cumulative effect of wetlands on stream water quality and quantity: a landscape approach. Biogeochemistry 10:105-141.

Lant, C.L., and G.A. Tobin. 1989. The economic value of riparian corridors in corn belt floodplains: a research perspective. Professional Geographer 41(3): 337-349.

Lawrie, A.H., and J.F. Rahrer. 1973. Lake Superior: a case history of the lake and its fisheries. Great Lakes Fishery Commission Technical Report 19. Great Lakes Fishery Commission, Ann Arbour, Michigan, USA.

Lisle, T.E. 1982. Roughness elements: a key resource to improve anadromous fish habitat. Pages 93-98 in T.J. Hassler, Editor. Proceedings on the propagation, enhancement, and rehabilitation of anadromous salmonid populations and habitat in the Pacific Northwest. California Cooperative Fisheries Research Unit,

Humboldt State University, Arcata, California, USA.

Long, B.A. 1987. Recruitment and abundance of large woody debris in an Oregon coastal stream system. Masters Thesis, Oregon State University, Corvallis, Oregon, USA.

Lowrance, R., R. Leonard, and J. Sheridan. 1985. Managing riparian ecosystems to control nonpoint pollution. Journal of Soil and Water Conservation **40**(1):87–91.

McNatt, R. 1978. Possible strategies for preservation of the San Padre River's riparian community. Pages 201–206 in R.R. Johnson and J.F. McCormick, editors. Strategies for protection and management of floodplain wetlands and other riparian ecosystems. United States Department of Agriculture Forest Service General Technical Report WO–12, Washington, D.C., USA.

Meehan, W.R., F.J. Swanson, and J.R. Sedell. 1977. Influences of riparian vegetation on aquatic ecosystems with particular references to salmonid fishes and their food supply. Pages 137–145 in R.R. Johnson and D.A. Jones, editors. Importance, preservation and management of riparian habitat: a symposium. United States Department of Agriculture Forest Service General Technical Report RM–43. Rocky Mountain Forest and Range Experiment Station, Fort Collins, Colorado, USA.

Megahan, W.F., and P.N. King. 1985. Identification of critical areas on forest lands for control of nonpoint sources of pollution. Environmental Management **9**(1):7–18.

Mormon, R.H., D.W. Duddy, and P.C. Rugen. 1980. Factors influencing the distribution of sea lamprey Petromyzon marinus in the Great Lakes. Canadian Journal of Fisheries and Aquatic Sciences **37**:1811–1826.

Nabhan, G.P. 1985. Riparian vegetation and indigenous southwestern agriculture: control of erosion, pests, and microclimate.

Pages 232–236 in R.R. Johnson, editor. Riparian ecosystems and their management: reconciling conflicting uses. United States Department of Agriculture Forest Service General Technical Report RM–120, Fort Collins, Colorado, USA.

Oregon State Department of Forestry. 1987. Streamside management concepts and recommendations. Salem, Oregon, USA.

Patterson, N.J., and T.H. Whillans. 1985. Human interference with natural water level regimes in the context of other cultural stresses on Great Lakes wetlands. Pages 209–251 in H.H. Prince and F.M. D'Itri, editors. Coastal wetlands. Lewis Publishers, Chelsae, Michigan, USA.

Peterjohn, W.T., and D.L. Correll. 1984. Nutrient dynamics in an agricultural watershed: observations on the role of a riparian forest. Ecology **65**:1466–1475.

Rainville, R.C., S.C. Rainville, and E.L. Linder. 1985. Riparian silvicultural strategies for fish habitat emphasis. Pages 186–196 in Technical Report 37. Proceedings of the 1985 National Convention. Society of American Foresters, Bethesda, Maryland, USA.

Regier, H.A., P. Tuunainen, Z. Russek, and L.E. Persson. 1988. Rehabilitative redevelopment of the fish and fisheries of the Baltic Sea and the Great Lakes. Ambio **17**:121–130.

Sedell, J.R., P.A. Bisson, F.J. Swanson, and S.V. Gregory. 1988. What we know about large trees that fall into streams and rivers. Pages 47–82 in C. Maser, R.F. Tarrant, J.M. Trappe, and J.F. Franklin, editors. United States Department of Agriculture Forest Service General Technical Report PNW–GTR–229. Pacific Northwest Forest and Range Experiment Station, Portland, Oregon, USA.

Sidle, R.C., A.J. Pearce, and C.L. O'Loughlin. 1985. Hillslope stability and land use. American Geophysical Union Water

Resources Monograph 11. Washington, D.C., USA.

Steedman, R.J., and H.A. Regier. 1987. Ecosystem science for the Great Lakes: perspectives on degradative and rehabilitative transformations. Canadian Journal of Fisheries and Aquatic Sciences **44**(2):95-103.

Swanson, F.J., L.E. Benda, S.H. Duncan, G.E. Grant, W.F. Magahan, L.M. Reid, and R.R. Ziemar. 1987. Mass failures and other processes of sediment production on Pacific Northwest Forest Landscapes. Pages 9-38 in E.O. Salo and T.W. Cundy, editors. Streamside management: forestry and fisheries interactions. University of Washington Institute of Forest Resources Contribution Number 57. University of Washington, Seattle, Washington, USA.

Swanson, F.J., S.V. Gregory, J.R. Sedell, and A.G. Campbell. 1982. Land-water interactions: the riparian zone. Pages 267-291 in R.L. Edmonds, editors. Analysis of coniferous forest ecosystems in the western United States. US/IBP Synthesis Series 14. Dowden, Hutchinson, and Ross Publishing Company, Stroudsburg, Pennsylvania, USA.

Swanston, D.N., editor. 1985. Proceedings of a workshop on slope stability: problems and solutions in forest management. United States Department of Agriculture Forest Service General Technical Report PNW-180. Pacific Northwest Forest and Range Experiment Station, Portland, Oregon, USA.

Ward, B.R., and P.A. Slaney. 1979. Evaluation of in-stream enhancement structures for the production of juvenile steelhead trout and coho salmon in the Keogh River: progress in 1977 and 1978. Fisheries Technical Circular 45. British Columbia Ministry of Environment, Victoria, Canada.

Whillans, T.H. 1984. Fish habitat alterations and present fishery management practices along the Canadian shore zone of the Laurentian Great Lakes. Pages 107-128 in Proceedings of non-salmonid rehabilitations workshop. Ontario Ministry of Natural Resources, Barrie, Ontario, Canada.

LANDSCAPE BOUNDARIES IN THE MANAGEMENT AND RESTORATION OF CHANGING ENVIRONMENTS: A SUMMARY

ROBERT J. NAIMAN AND HENRI DÉCAMPS. Center for Streamside Studies, AR-10, University of Washington, Seattle, Washington 98195, USA , and Centre National de la Recherche Scientifique, Centre d'Ecologie, 29 rue Jeanne Marvig, 31055 Toulouse France.

Abstract. The main challenges facing ecologists in the next decade are the creation of a landscape perspective among scientists, managers and policy-makers; determining the cumulative effects of land use practices; rehabilitating damaged ecosystems; designing management systems to accommodate natural variability; prudent conservation of natural systems; and clarifying relationships between human social and economic systems and the environment. In this chapter, we summarize key points addressing the role of ecotones (boundaries) for meeting these challenges in a changing global environment. We present fundamental research issues related to ecotones, lessons from this and six recent international symposia, and recommendations for advancing the applicability of the ecotone concept through research and education. We suggest that the main challenges for human societies are to maintain continued economic strength through wise and efficient use of existing natural resources, and to improve the quality of life while maintaining the quality of the environment, and that an ecotone perspective is one of several allied concepts that can be used to meet those related challenges.

Key words: Boundary, ecotone, environment, management, resource economics, society.

INTRODUCTION

During the next decade and into the next century, exploitation and modification of the earth's natural resources will increase, thus demanding enhanced management approaches (di Castri et al. 1984, di Castri and Hadley 1988). These enhanced management approaches will require an understanding about how ecosystems operate over longer time periods, especially with respect to the cumulative effects of land use practices. Just as important will be the recognition that landscapes consist of interacting ecosystems. Between these ecological units are boundaries, or ecotones, that affect the behavior of the landscape as a whole.

As described in Naiman and Décamps (1990*b*), there are at least six main challenges facing ecologists at the ecosystem or landscape level. These are:

1.   The creation of a landscape perspective among scientists, managers and policy-makers;

2.   Determining the cumulative effects of land use practices;

3.   Rehabilitating damaged ecosystems;

4.   Designing management systems to accommodate natural variability;

5.   Prudent conservation of natural systems; and,

6.   Clarifying relationships between human social and economic systems and the environment.

These challenges can be partly met with an ecotone-patch dynamics approach. For example, considering the six main challenges separately:

1.   A landscape perspective requires the treatment of large areas as a mosaic of interacting patches partly regulated by their ecotones,

2.   Cumulative effects result from perturbations being transmitted through the system along pathways that are often unpredictable and irreversible. Ecotones act to modify, or even control, transmission of disturbances between patches.

3.   Ecotones can also be used to assist

130

the recovery of patches from disturbance by using their natural properties for creative solutions.

4. Ecotones, patches, and ecosystems are naturally variable in time and space. Yet most ecosystems are managed for stability. An ecotone approach would require a new, and needed, perspective for the management of uncertainty.

5. The difficult issues related to the conservation of natural systems could be resolved more easily if key ecotonal features were preserved and protected.

6. Finally, many of the environmental issues we face are directly driven by global resource economics, international monetary exchange rates and social dynamics. Economics influences land use patterns and environmental quality over long distances through the value and demand for natural resources. A patch-ecotone perspective will be required if we are to integrate environmental issues with economic and social driving forces.

These challenges become increasingly important when one considers that the environment represents an essential element for the social, economic, and ecological well-being of every country on earth. Terrestrial and aquatic resources have multiple uses for agriculture, industry, recreation, transport, fisheries, environmental functions, and human health. Yet, our natural environment is increasingly in need of better protection and management, and ecotones have been identified as key components affecting, and perhaps regulating, significant aspects of the overall environmental quality (Risser 1985, di Castri et al. 1988, Lauga et al. 1988, Naiman and Décamps 1990a,b).

LESSONS FROM OTHER SYMPOSIA

Although the definition of ecotones contains the understanding of transitions between ecological systems, it must be recalled that the identification of a particular ecotone depends on the question being asked or the problem to be solved. Ecological questions are addressed over spatial scales ranging from 10^{-8} to 10^{7}m and temporal scales from 10^{-7} to 10^{8} years, or sixteen orders of magnitude (Minshall 1988). Over these spatial and temporal scales, the theory of ecosystems is quite immature (Naiman et al. 1988a, b, 1989, Shugart 1990), although it is clearly recognized that the spatial and temporal arrangements of resource patches and their associated ecotones are functionally important (Pickett and White 1985, Forman and Godron 1986, Pringle et al. 1988).

Ecotones are created and maintained by external environmental forces that correspond to the scale at which the ecotone is identified. These forcing functions (e.g. disturbance, edaphic factors, climate, geomorphology) can be recognized at different scales, and as a result, investigations of ecotones must focus at the identified scale (Naiman et al. 1989). Ecotones, because of their special characteristics, may be sensitive to changes in these forcing functions. For example, ecotones might be quite sensitive to global climate change although the actual responses will depend upon the nature of the specific ecotone. Similarly, depending upon the external forcing functions, or the weaker internal mechanisms, ecotones may exhibit higher or lower levels of biodiversity (di Castri et al. 1988). Similarly, ecotones may modify the flows of materials moving through these transitional zones, depending both on the external force, and their internal characteristics (Correll 1986, Pinay et al. 1990).

The management of ecotones depends on understanding these fundamental properties, but also incorporating social and economic considerations into a "systems approach." Because of the resulting complexity, certain management approaches have been developed, e.g., Best Possible Management Option (BPMO) in England and the Timber, Fish, Wildlife Agreement (TFW) in Washington State (Halbert and Lee 1990, Petts 1990). Curiously, much of our previous management philosophy has been based on environmental stability, whereas we now know that aquatic-terrestrial ecotones are characterized by instability and variability.

LESSONS FROM THIS SYMPOSIUM

The presentations in this symposium had, understandably, some overlap with presentations at previous symposia. Since 1986 there have been six international conferences convened, including this one, to discuss the validity of the ecotone concept, to evaluate the state-of-knowledge, and to make recommendations as to how the concept could be applied to societal issues related to environmental change (Holland and Risser this volume). However, many similar conclusions were reached through largely independent thought processes and experiences, giving the statements made in the previous section more credence.

James R. Gosz reviewed fundamental ecosystem characteristics of landscape boundaries noting that many of the research issues (such as scale, scale dependent results, assumptions of methods, and technique dependent results) transcend all areas of ecology. However, it was also noted that advances are being made in boundary detection with new applications of mathematical techniques which allow identification of boundaries at the biotic scale of the organism rather than only the human scale. He effectively argues for the urgent need for new techniques and approaches which encompass a landscape perspective as well as multiple and dynamic scales.

Ronald P. Neilson examined the relation between climate and scale in controlling ecotones located between major regional biomes. The concept that ecotones can be potentially sensitive features for monitoring impending climate change was explored from the perspective of detecting ecotone location, and change in location, using regional gradients in landscape patterns of habitat variability. Two types of change (e.g., boundary change and within-region quality change) were viewed

as being potentially independent. As such they may require different monitoring strategies to detect impending change. Neilson also addressed processes that appear to control both coarse and fine patterns in habitat diversity suggesting that, if they can be identified, they then become early warning signals of global change. Likewise, Neilson hypothesized that if spatial and temporal gradients can be related to common mechanisms of habitat structure, we may be able to apply rules gained from analyzing the spatial patterns to the development of early warning signals of temporal environmental change. Finally, he demonstrates that all major biomes in the United States appear to be produced by well defined gradients in seasonal weather patterns, or by threshold responses at critical points along thermal gradients, which can be used to evaluate longer-term climate alterations.

Monica G. Turner and her colleagues used neutral models to examine how altered disturbance regimes and habitat features may affect the location, size, shape and composition of landscape boundaries. They found that, in connected landscapes, an increase in disturbance intensity caused landscape boundaries to disappear. Whereas, in fragmented landscapes, increase in disturbance frequency was responsible for boundary disappearance. When ecotonal habitat was displaced, the biotic community could track habitat movement when the displacement rate was slow, when the probability of extinction was low, and when the chance to migrate was moderate. They were able to demonstrate the utility of neutral models to generate quantitative, testable hypotheses which could be used to understand and predict landscape responses to global change.

Robert H. Gardner and his colleagues used a simulation model to examine the movement of organisms with contrasting life history characteristics and abundances through heterogeneous landscapes with different boundary characteristics and densities. They found that the scale at which a species can move in an environment with patchy resources is determined by its inherent dispersal mechanisms. Their simulations suggest that where the fraction of available habitat is less than the 0.6 threshold predicted by percolation theory, the spread of organisms across landscapes is sensitive to differences in life history characteristics and dispersal mechanisms between species. In this case, boundaries become important. Above 0.6 all patches are well connected and do not generally impede individual species, and landscape boundaries that would regulate species movement largely cease to exist. They noted that the existence of a critical threshold is important for many management and conservation issues, but the determination of these thresholds will require data gathered at scales that are relevant to each species.

David L. Correll applied the ecotone concept to intensely managed agricultural-forested landscapes, noting that human management often creates or sharpens boundaries between natural and managed systems, and that these boundaries often correspond to natural (i.e., geomorphic) landscape features. He demonstrated that riparian forest ecotones have an important role in regulating and transforming nutrients moving from agricultural lands to streams. The management implications of this work are enormous, especially as the fragmentation of the world's riparian forests continue. Both the ecological and economic values of the riparian ecotone are large, providing practical reasons for maintaining and enhancing riparian vegetation. Unfortunately, only a few of the supposedly numerous ecological functions of riparian forests have been explored. At the same time, these ecotones could be among the first to respond to climate change since their inherent characteristics are so closely regulated by the hydrologic regime.

James R. Sedell and his colleagues provide several examples for successful restoration of human impacted ecotones.

They point to examples in the Pacific Northwest, the Arid West, and the Laurentian Great Lakes, where riparian and nearshore areas function as especially valuable areas for fish and wildlife, and for commercial or aesthetic considerations. Because of the importance of these areas for a wide variety of values (i.e., water quality and quantity, aesthetics, fisheries, wildlife, forage, timber, and recreation), the general public is increasingly involved in riparian and wetland regulation and policy decisions. The development of informed environmental constituencies must precede the coordinated movement of hundreds of local, state or provincial, national, and international regulatory jurisdictions toward common environmental goals. Such a change clearly depends on formulation of ecologically relevant natural resource and economic policies. These developments must be guided by ecosystem science that fosters regional understanding of ecosystem function.

NEEDS FOR THE FUTURE

As described in Naiman and Décamps (1990b), there are two major categories of needs for the advancement of an 'ecotone perspective.' These are an improved knowledge base and development of new tools and techniques. Each represents a major challenge for effective implementation.

We suggest that the major improvements needed for the knowledge base can be grouped into four themes:

1. Synthetic models of complex phenomena and complex systems. Environmental issues are becoming increasingly difficult to resolve because of their spatial extent and their impact on social-economic groups which possess contrasting interests or perspectives. Natural resource scientists are being asked to understand phenomena that are often beyond the capabilities of a single human mind. The resolution of these issues requires an interdisciplinary approach that has here-

to-fore been used in the United States space program and in the operation of successful corporations, and is envisioned as the basis for the operation of the United States Global Change Program (see Holland and Risser this volume).

2.   A classification scheme based on appropriate spatial and temporal scales. As was effectively articulated in these proceedings, the scale of the issue dictates the scale of the research perspective. It is, at best, difficult to extrapolate from small scale observations to global issues. Similarity, global experimentation is not feasible. How to move reliability between scales, therefore, becomes an essential task to be mastered.

3.   The use of ecotones as natural systems to speed the recovery of damaged systems. The use of ecotonal features to aid the restoration of damaged systems can be partially accomplished with existing knowledge. However, it will require adapting existing and emerging knowledge from other disciplines to the problem at hand. Effective technology transfer is an important consideration in attempting to rebuild the functional and structural integrity of damaged landscapes. An example is using silvicultural techniques and genetic selection processes to restore the ecotonal features of riparian forests which influence downstream water quality, wildlife diversity and fish production (Naiman et al. *in press*).

4.   Mechanisms for managerial implementation and societal awareness of ecotonal values. A key process in effective management is to insure that there is open and well-articulated communication between scientists, managers, and resource consumers. In many ways, management is an experimental science which can be used to test ideas about landscape and societal processes. This concept is central to the adaptive management philosophy proposed by Holling (1978), Walters (1986) and Lee (1989). Yet most regions have poorly developed mechanisms and processes for the effective implementation of new information. Further, since research programs

cannot be entirely designed in advance, an essential element is the flexibility to change and redesign immediate objectives while keeping the final objective the same. Both are vital links in the adaptive management process. Addressing these knowledge gaps can be partly accomplished with existing tools and techniques. However, we stress the need to advance quickly in developing new techniques and in adapting existing methods from other disciplines.

We suggest the major tools and techniques also can be grouped into four broad themes:

1.   Improved training and use of high speed computers for evaluating connectivity, variability, and cumulative effects in complex systems. Environmental scientists are being asked to understand and respond to increasingly complex and often intractable issues. Access to high speed computers and the modeling expertise however, are difficult to obtain for natural resource based issues. Access and training at all educational levels are urgently needed, if the emerging issues of the next decade are to be effectively addressed.

2.   Development of spatial and temporal statistical methods for the quantitative evaluation of patch and boundary dynamics. Traditional mathematical and statistical techniques are not adequate for evaluating relations between landscape dynamics and ecological processes (such as factors important in determining water quality). New techniques are essential for identifying factors responsible for environmental quality and productivity.

3.   Continued development of techniques associated with Geographic Information Systems (GIS), nuclear magnetic resonance (NMR), and remote sensing (SPOT, Landsat, EOS) that have direct relevance to ecotone identification and processes. These tools have the potential to advance ecological science to scales that are appropriate to the issues at hand. As years pass, the accessibility of these tools increases,

promising improved understanding of fundamental processes over large areas. However, accelerated training is required and new tools as yet undiscovered need to be adapted for ecological uses to keep pace with the issues that natural resource scientists must address.

4. Development of long-term demonstration sites for interdisciplinary and cross-cultural education at all levels (e.g., societal, scientific and managerial). Certainly the National Science Foundation's Long Term Ecological Research sites have provided much valuable information about functioning ecosystems in the United States. However, even more sites need to be established for the orderly and efficient training of large numbers of individuals encompassing a full array of scientific and cultural philosophies. Many of these needs are being addressed independently in other programs concerned with environmental issues in the next decade. A major task will be to adapt the knowledge, the tools, and the techniques which will contribute to the development of an 'ecotonal perspective' and to the resolution of issues. If successfully adapted, these improvements could provide a solid conceptual and practical framework for improving environmental understanding, as well as the human condition, by providing for diversity, resilience, productivity, and sustainability throughout the landscape.

CONCLUSION

As we approach the 21st century, the main challenge for humans has shifted away from developing societal and economic systems exclusively based on natural resource exploitation. The positive economic aspects of an 'ecotone perspective' include long term sustainable systems, a better understanding of factors controlling variability in natural resources, a perspective that rewards stewardship rather than exploitation, and reduced conflict between nations seeking common resources. If ecotones are found to be key landscape elements, then these economic aspects present a strong argument for considering an ecotone perspective for ecological systems.

ACKNOWLEDGEMENTS

We thank D.L. Correll, J. Gosz, M.M. Holland, M.N. Jensen, R. Lee, B. Milne, J. Pastor, P.G. Risser, J.R. Sedell, and an anonymous reviewer for comments on the manuscript. Financial support was provided by the National Science Foundation (BSR 87-22852), the Centre National de la Recherche Scientifique (France), the United States Man and the Biosphere Program (Department of State) and the UNESCO Man and the Biosphere Programme.

LITERATURE CITED

Correll, D.L., editor. 1986. Watershed research perspectives. Smithsonian Institution Press, Washington, D.C., USA.

di Castri, F., W.G. Baker, and M. Hadley, editors. 1984. Ecology in practice. Tycooly International Publishing, Dublin, Ireland.

di Castri, F., and M. Hadley. 1988. Enhancing the credibility of ecology: interacting along and across hierarchical scales. Geo Journal **17**:5-35.

di Castri, F., A.J. Hansen, and M.M. Holland, editors. 1988. A new look at ecotones: emerging international projects on landscape boundaries. Biology International, Special Issue **17**:1-163.

Forman, R.T.T., and M. Godron. 1986. Landscape ecology. John Wiley and Sons, New York, New York, USA.

Halbert, C.L. and K.N. Lee. 1990. The Timber, Fish and Wildlife Agreement: implementing alternative dispute resolution in Washington State. Northwest Environmental Journal **6**:139-175.

Holland, M.M., and P.G. Risser. 1991. The role of landscape boundaries in the management and restoration of changing environments: introduction, this volume.

Holling, C.F. editor. 1978. Adaptive environmental assessment and management. John Wiley and Sons, New York, New York, USA.

Lauga, J., H. Décamps, and M.M. Holland, editors. 1988. Land use impacts on aquatic ecosystems: the use of scientific information. Académie de Toulouse, Toulouse, France.

Lee, K.N. 1989. The Columbia River Basin: experimenting with sustainability. Environment 31:7–33.

Minshall, G.W. 1988. Stream ecosystem theory: a global perspective. Journal of the North American Benthological Society 7:263–288.

Naiman, R.J., H. Décamps, J. Pastor, and C.A. Johnson. 1988a. The potential importance of boundaries to fluvial ecosystems. Journal of the North American Benthological Society 8:289–306.

Naiman, R.J., M.M. Holland, H. Décamps, and P.G. Risser. 1988b. A new UNESCO programme: research and management of land/inland water ecotones. Biology International, Special Issue 17:107–136.

Naiman, R.J., H. Décamps, and F. Fournier, editors. 1989. The role of land/inland water ecotones in landscape management and restoration. Man and the Biosphere Digest 4. UNESCO, Paris, France.

Naiman, R.J., and H. Décamps, editors. 1990a. Ecology and Management of Aquatic-Terrestrial Ecotones. Man and the Biosphere Series. The Parthenon Publishing Group, Carnforth, United Kingdom.

Naiman, R.J., and H. Décamps. 1990b. Aquatic-terrestrial ecotones: summary and recommendations. Pages 295–303 in R.J. Naiman and H. Décamps, editors. Ecology and management of aquatic-terrestrial ecotones. Man and the Biosphere Series. The Parthenon Publishing Group, Carnforth, United Kingdom.

Naiman, R.J., D.G. Lonzarich, T.J. Beechie, J.C. Ralph. In press. Stream classification and the assessment of conservation potential. In P. Boon, editor. Conservation and management of rivers. Wiley and Sons, England.

Petts, G.E. 1990. The role of ecotones in aquatic landscape management. Pages 227–262 in R.J. Naiman and H. Décamps, editors. The ecology and management of aquatic-terrestrial ecotones. Man and the Biosphere Series. The Parthenon Publishing Group, Carnforth, United Kingdom.

Pickett, S.T.A., and P.S. White, editors. 1985. The ecology of natural disturbance and patch dynamics. Academic Press, New York, New York, USA.

Pinay, G., H. Décamps, E. Chauvet, and E. Fustec. 1990. Functions of ecotones in fluvial systems. Pages 141–170 in R.J. Naiman and H. Décamps, editors. The ecology and management of aquatic-terrestrial ecotones. Man and the Biosphere Series. The Parthenon Publishing Group, Carnforth, United Kingdom.

Pringle, C.M., R.J. Naiman, G. Bretschko, J.R. Karr, M.W. Oswood, J.R. Webster, R.L. Welcomme, and M.J. Winterbourn. 1988. Patch dynamics in lotic systems: the stream as a mosaic. Journal of the North American Benthological Society 7:503–524.

Risser, P.G., compiler. 1985. Spatial and temporal variability of biospheric and geospheric processes: research needed to determine interactions with global environmental change. Report of a workshop sponsored by SCOPE, INTECOL, and ICSU. 18 October – 1 November 1985. St. Petersburg, Florida, USA.

Shugart, H.H. 1990. Ecological models and the ecotone. Pages 23–36 *in* R.J. Naiman and H. Décamps, editors. The ecology and management of aquatic–terrestrial ecotones. Man and the Biosphere Series.

The Parthenon Publishing Group, Carnforth, United Kingdom.

Walters, C. 1986. Adaptive management of renewable resources. Macmillan Company, New York, New York, USA.

INDEX